Fundamentals of Robotics

MECHANICAL ENGINEERING

A Series of Textbooks and Reference Books

EDITORS

L. L. FAULKNER

*Columbus Division
Battelle Memorial Institute*

and

*Department of Mechanical Engineering
The Ohio State University
Columbus, Ohio*

S. B. MENKES

*Department of Mechanical Engineering
The City College of the
City University of New York
New York, New York*

1. Spring Designer's Handbook, *by Harold Carlson*
2. Computer-Aided Graphics and Design, *by Daniel L. Ryan*
3. Lubrication Fundamentals, *by J. George Wills*
4. Solar Engineering for Domestic Buildings, *by William A. Himmelman*
5. Applied Engineering Mechanics: Statics and Dynamics, *by G. Boothroyd and C. Poli*
6. Centrifugal Pump Clinic, *by Igor J. Karassik*
7. Computer-Aided Kinetics for Machine Design, *by Daniel L. Ryan*
8. Plastics Products Design Handbook, Part A: Materials and Components; Part B: Processes and Design for Processes, *edited by Edward Miller*
9. Turbomachinery: Basic Theory and Applications, *by Earl Logan, Jr.*
10. Vibrations of Shells and Plates, *by Werner Soedel*
11. Flat and Corrugated Diaphragm Design Handbook, *by Mario Di Giovanni*
12. Practical Stress Analysis in Engineering Design, *by Alexander Blake*
13. An Introduction to the Design and Behavior of Bolted Joints, *by John H. Bickford*
14. Optimal Engineering Design: Principles and Applications, *by James N. Siddall*
15. Spring Manufacturing Handbook, *by Harold Carlson*
16. Industrial Noise Control: Fundamentals and Applications, *edited by Lewis H. Bell*
17. Gears and Their Vibration: A Basic Approach to Understanding Gear Noise, *by J. Derek Smith*

Mechanical Engineering Software

Fundamentals of Robotics

David D. Ardayfio

Department of Mechanical Engineering
Wayne State University
Detroit, Michigan

CRC Press
Taylor & Francis Group
Boca Raton London New York

CRC Press is an imprint of the
Taylor & Francis Group, an **informa** business

First published 1987 by Marcel Dekker , Inc.

Published 2019 by CRC Press
Taylor & Francis Group
6000 Broken Sound Parkway NW, Suite 300
Boca Raton, FL 33487-2742

©1987 by Taylor & Francis Group, LLC
CRC Press is an imprint of Taylor & Francis Group, an Informa business

First issued in paperback 2019

No claim to original U.S. Government works

ISBN-13: 978-0-367-45147-9 (pbk)
ISBN-13: 978-0-8247-7440-0 (hbk)

Visit the Taylor & Francis Web site at
http://www.taylorandfrancis.com

and the CRC Press Web site at
http://www.crcpress.com

Library of Congress Cataloging-in-Publication Data

Ardayfio, David D.
   Fundamentals of robotics.

   (Mechanical Engineering ; 57)
   Includes bibliographies and index.
   1. Robotics. I. Title. II. Series.
TJ211.A73   1987   629.8'92   87-6850
ISBN 0-8247-7440-X

Preface

This text is an attempt to bring together several concepts in the interdisciplinary area of robotics. Early in 1982, the author was assigned to teach an introductory overview college course in robotics to senior mechanical engineering students. At the time no one book could be found that contained all the basic qualitative and quantitative knowledge on robots.

The objective of this book is to present the basic concepts of robots to engineering and technology students and to practicing engineers who want to grasp the fundamentals in the growing field of robotics.

The text is organized into twelve chapters. Chapter 1 gives an overview of robots: definitions, basic robot types, actuators, and transmissions. Examples of industrial robots are also presented to reflect the basic design concepts. Chapter 2 treats the definition, types, and design considerations in the application of grippers. The next chapter is devoted to the discussion of a number of end-of-arm tooling components. Chapter 4 deals with end-of-arm accessories having tactile sensors and compliant wrist devices. In Chapter 5, we introduce the notion of robot control from simple fixed sequence mechanical controllers to sophisticated computer controllers. Chapter 6 deals with nontextual programming of robots. This is then extended to textual programming using computer languages in Chapter 7. Robot vision systems are described in Chapter 8. Major applications of robots in manufacturing operations are discussed in Chapter 9.

The next chapters present the analytical basis for robots covering the following topics: workspace geometry, kinematics force analysis, trajectory generation, and control.

David D. Ardayfio

Contents

Introduction

In examining what a robot is, there are various definitions and various interpretations of those definitions. The Robot Institute of America defines a robot as a programmable, multifunction manipulator designed to move materials, parts, tools, or specialized devices through variable programmed motions for the performance of a variety of tasks. The key words in this definition are reprogrammable and variety. Computer-Aided Manufacturing International offers the definition of a robot as a device that performs functions ordinarily ascribed to human beings or that operates with what appears to be almost human intelligence. The International Standards Organization describes a robot as a machine formed by a mechanism including several degrees of freedom, often having the appearance of one or several arms ending in a wrist capable of holding a tool, a workpiece, or an inspection device. Further, the control unit of the robot must use a memorizing device and sometimes can use sensing and adaptation appliances, taking into account environment and circumstances. These multipurpose machines are generally designed to carry out a repetitive function and can be adapted to other functions.

The common concept in these three descriptions of a robot is that it should be a type of industrial device that is as general-purpose as perhaps the workers it might be replacing. This concept rules out special-purpose or hard automation as a robot because it is built for a single purpose and rarely adaptable for another job. These definitions would also exclude remote manipulators which are extensions of a human operator's hands.

The Japanese Industrial Robots Association defines a robot according to the modes of control as (1) fixed sequence, (2) variable

sequence, (3) playback, (4) numerical control, and (5) intelligent.
Some devices that might therefore be counted as robots by the Jap-
anese definition are not robots according to the Robot Institute of
America.

At the end of 1981, an estimated 14,000 robots were in operation
in Japan, 4,400 in Western Europe, 3,000 in the USSR, and 4,100
in the United States. At the Twelfth International Symposium on
Industrial Robots in 1982, the following numbers were given:
Japan, 6,100; United States, 3,500; Sweden, 1,200; West Germany,
1,130; and Italy, 400. The discrepancies in any estimated robot
population are due largely to the various ways of defining a robot.

Fundamentals of Robotics

1

Basic Components of Robot Arms

1.1 BASIC ARM TYPES

The identification of the various kinematic arrangements used in industrial robots is useful for manufacturers, users, and researchers. The working volume and dexterity, for example, of a robot are determined by the kinematic arrangements. From the user's point of view, a preliminary determination of the performance of a robot in the work environment based solely on geometry is possible by examination of the linkage type forming the structure of the robot.

Manipulators are typically located relative to ground by either a fixed base or a moving base. The first three links of the manipulator collectively form the manipulator arm. This arm is used for global positioning of the end of the manipulator within the work envelope. The hand is composed of the other links (up to three) and is used to define the orientation of the end of the manipulator. The hand also provides fine local positioning of the end of the manipulator. Detailed treatment of the hands and grippers used in manipulators can be found in Chapter 2. The present chapter discusses various arm types and other components of robot setups.

Two of the three types of connecting elements with one degree of freedom that can be used for connecting links in a robot are the prismatic and revolute pairs. The prismatic or sliding pair allows pure translation of one link with respect to another. The revolute or rotation pair provides pure rotation of one link on another.

The structural arrangement of robot linkage systems (Figure 1.1) includes:

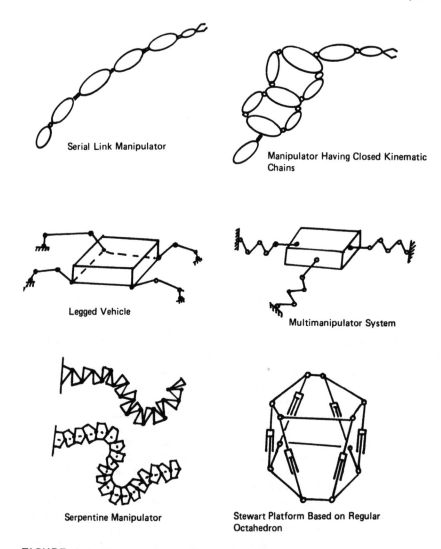

Serial Link Manipulator

Manipulator Having Closed Kinematic Chains

Legged Vehicle

Multimanipulator System

Serpentine Manipulator

Stewart Platform Based on Regular Octahedron

FIGURE 1.1 Types of robotic mechanisms.

1. Robot manipulators with simple serial kinematic chains
2. Robot manipulators with tree structures
3. Robot manipulators containing closed kinematic chains
4. Legged locomotory vehicles
5. Multimanipulator systems and dextrous manipulator grippers
6. Parallel connection arrangements
7. Multisegmented serpentine designs

Serial link robots (Figures 1.2 to 1.9) may be classified according to type and sequence of the kinematic arrangements of links with prismatic (P) and revolute (R) joints in four main categories:

1. Cartesian ($P_1P_2P_3$)
2. Cylindrical ($R_1P_2P_3$)
3. Spherical ($R_1R_2P_3$)
4. Anthropomorphic ($R_1R_2R_3$)

A cartesian robot, which is also called rectangular or rectilinear, has three translation joints. It generates a rectangular workspace with the three basic motions: base travel (P_1), reach (P_2) and elevation (P_3). Some typical industrial robots of this classification are:

Source	Model	Drive system
Advanced Robotics (USA)	Cyro 750	Electric DC
Anorad (USA)	Anorobot	Electric DC
FATA European Group (Italy)	Junior	Electric
	Senior	Electric
Hitachi (Japan)	Mr: Aros	Hydraulic
Matsushita Ind. (Japan)	RW 2000	Electric AC
Mitsubishi Heavy Ind. (Japan)	Robitus RC-RH	Hydraulic
Mouldmation (UK)	Mouldemate 1200	Pneumatic
Oy W Roseniew (Finland)	Rb 17, 19	Hydraulic
Renault (France)	G 80	Hydraulic
S.A. Distribel Robotechnic (Belgium)	Phoenix 3512S	Electric stepper
Volkswagenwerke AG (West Germany)	R 100	Electric DC
VFW (West Germany)	E 440	Electric DC

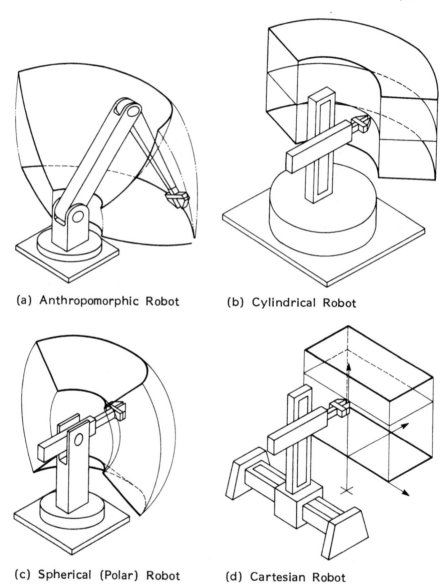

(a) Anthropomorphic Robot   (b) Cylindrical Robot

(c) Spherical (Polar) Robot   (d) Cartesian Robot

FIGURE 1.2 Basic configuration of serial robot arms and the associated workspace.

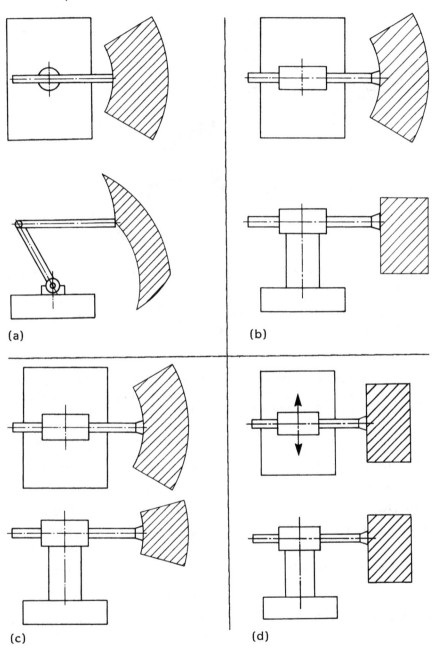

(a)

(b)

(c)

(d)

FIGURE 1.3   Orthographic projections of the workspace of the basic robot configurations.   (a) Anthropomorphic robot, (b) Cylindrical robot, (c) Polar robot, (d) Rectangular robot.

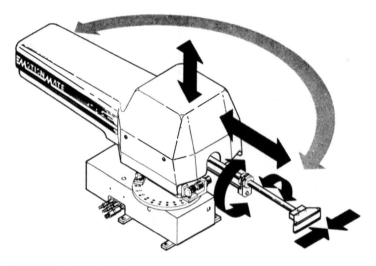

FIGURE 1.4 Configuration of a robot with cylindrical coordinate arm motions. (A) Base rotates with a motion range of 180°, which is adjustable in increments of 15°; (B) lift of 3 in.; (C) extends to 12 in.; (D) wrist rotates 90° or 180°; (E) grasp. (Courtesy of Schrader-Bellows, a Division of Scovil.)

The cylindrical robot (Figure 1.4) has its basic motions in cylindrical coordinates: base rotation (R_1), reach (R_2), and elevation (P_3). Examples of cylindrical robots are:

Source	Model	Drive system
FAB Bilsing (Germany)	High speed	Hydraulic
Fanuc (Japan)	M0, M1, M2	Electric
Jobs (Italy)	JOBOT 10	Electric
Kayaba Ind. (Japan)	KMR 200	Hydraulic
	KMR 300	Hydraulic
Lamberton Robotics (UK)	AA 150	Electric
	AA 700	Electric
	AA 1300	Electric
Oy W Resenjew (Finland)	RB 14	Hydraulic
	RB 15	Hydraulic
Shinko Electric (Japan)	SR-25	Pneumatic
Sterling Detroit (USA)	Robotarm	Hydraulic
VFW (West Germany)	E404	Electric

FIGURE 1.5 Four geometric configurations of a robot. (Courtesy of PICKOMATIC Systems.)

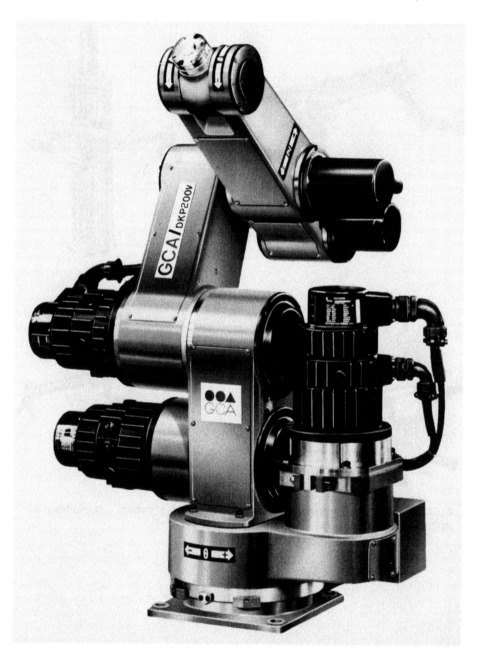

FIGURE 1.6 An anthropomorphic robot with joint offsets. (Courtesy of GCA Corporation.)

FIGURE 1.7 Prosthetic configuration of the GCA/DKP200H robot.
(Courtesy of GCA Corporation.)

In the spherical or polar arm configuration, a spherical workspace in generated by base rotation (R_1), elevation angle (R_2),
and reach (P_3). Some industrial robots of this geometric type are:

Manufacturer	Model	Drive system
Armax Robotics Inc. (USA)	LC/VC	Hydraulic
Comau Industriale (Italy)	Polar 6000	Hydraulic

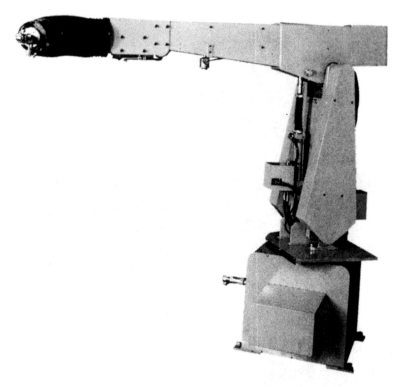

FIGURE 1.8 A painting robot with flexible wrist. (Courtesy of
Tokico America, Inc.)

Manufacturer	Model	Drive system
Fairey Automation (UK)	Automator	Hydraulic
GEC Robot Systems (UK)	Little Giant	Hydraulic
Prab Robots (USA)	4200 HD	Hydraulic
	5800 HD	Hydraulic
Toshiba Seiki (Japan)	IX-155	Hydraulic
Unimation (USA)	2000, 4000	Hydraulic
Volkswagenwerke (Germany)	R 30	Electric DC

An anthropomorphic robot is an articulated arm with motions
similar to human movements. An anthropomorphic arm is revolute

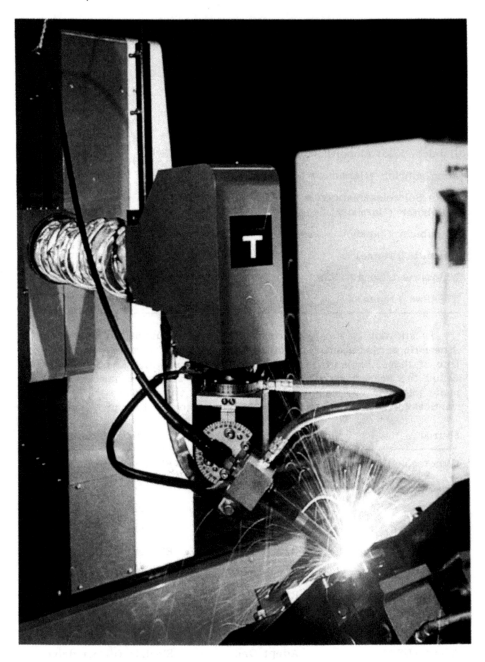

FIGURE 1.9 An arc welding robot with selectable welding positions.
(Courtesy of Merrick Engineering, Inc.)

jointed, with or without joint offsets, capable of having base (waist) rotation (R_1), shoulder rotation (R_2), and elbow rotation (R_3). Examples of this type of robot are:

Source	Model	Drive system
Bendix Robotics Div. (USA)	ML-360	Electric DC
Cincinnati Milacron (USA)	T^3 566	Hydraulic
Dainichi Kiko (Japan)	BABOT 4700	Electric DC
Jungheinrich (Germany)	A100	Electric DC
Kuka Schweissanlangen & Roboter (Germany)	160/60	Electric DC
Mitsubishi (Japan)	Robitus RD	Hydraulic
Renault (France)	V80, H80	Hydraulic
Thermwood Corp. (USA)	Series 3	Hydraulic
Yaskawa (Japan)	S-30	Electric

In addition to the four common types of industrial robots, other kinematic arrangements and other types of robot structures exist. One of these other types is the prosthetic arm which is commonly called the Scara robot. The prosthetic arm structure has a prismatic joint for vertical positioning (P_1) and two revolute joints with vertical axes (R_2, R_3) for the horizontal sweep. Examples include:

Source	Model	Drive system
Cybotech (USA)	H680	Electric/hydraulic
GCA (USA)	DKP200H	Electric
DeVilbiss (USA)	EPR 400, 405	Electric
IBM (USA)	IBM 7535, 7540	Electric
Motoman (Japan)	S50	Electric
Robomatic (USA)	700 Series	Mechanical
Mitsubishi (Japan)	RH-211	Electric
Hirata (Japan)	AR H300 300	Electric
Adept (USA)	Adept One	Electric (direct drive)
Accusembler (USA)	SSR-4	Electric
Anorad (USA)	AnoArm	Electric

Other robot structures have closed linkage arrangements with four-bar or five-bar mechanisms. A simple parallelogram manipulator arm (Figure 1.10) is obtained from the pantograph mechanism used frequently in drafting for scaling drawings. Pantograph mechanisms may be constructed as robot arms in either cylindrical or rectangular coordinates.

Two parallel kinematic chains may be connected by links with three or four connection points to form a manipulator arm as shown in Figure 1.11. Articulated manipulator arms used to locate tools must be able to move in a straight line. Parallelogram linkage robots are simple arm structures for such required rectilinear motion of the end effector. Fcr horizontal straight-line motion of the hand, both parallelograms may be folded or extended. The vertical column supporting the rear end of the first parallelogram may also be turned on a swivel. The triangular form of the connecting link is the simplest, but the double-revolute joint on the common point of both parallelograms may present practical difficulties in the design.

One of the advantages of linkage robots is that a path which is generated by a mechanism with two degrees of freedom can usually be done with a constrained four-bar linkage. This reduces the number of actuators with a consequent simplification of the robot control. Generally, linkage mechanism robots tend to be lightweight in construction and can handle heavier loads with reduced power consumption

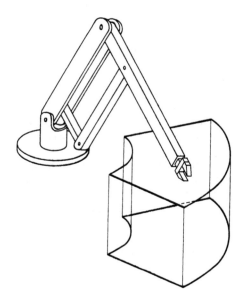

FIGURE 1.10 Workspace of a Pantographic parallelogram robot arm.

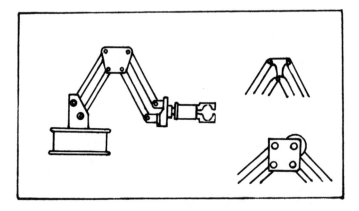

FIGURE 1.11 Structure of parallelogram linkages connected by ternary or quaternary links to form a robot arm.

However, errors in linkage mechanisms limit the positioning accuracy of the robot.

In a pendulum (sometimes called portal) configuration, a robot is mounted as a swinging arm which hangs like a pendulum that concentrates the mass of the robot at the pivot point. This above-floor suspension arrangement allows up to 50% faster acceleration than conventional arrangements.

Gantry robots, which are becoming increasingly popular, belong to this classification. These robots are ideal for transferring parts into spaces with limited floor access and for lifting heavy weights. The gantry layout has the added advantage of allowing easy access for tool setting and maintenance.

The overhead gantry robot shown in Figure 1.12 is used in, for example, nuclear power plants, aluminum mills, and the mining industry. Direct electric drives on the carriage and bridge (X and Y) axes with a dual resolver system enhance the positional accuracy of this robot. Rack-and-opinion gearing on the X and Y axes may also be used.

A nonrigid serpentine structure painting industrial robot consists of a number of discs held together by pretensioned cables that give strength, stability, and precision. The arm moves as the forces acting on the cables vary. The flexible arm sections give a wide range of movement to maneuver freely and get in and out of places normally considered inaccessible to robots.

Mobile robots (Figure 1.13) have the ability to run on rails or move on the floor. Some mobile robots can go up and down stairs, go over obstacles, and turn in narrow spaces.

FIGURE 1.12 A gantry-mounted robot.   (Courtesy of GCA/PaR
systems.)

FIGURE 1.13 Welding robot capable of moving on rails for large workpieces. (Courtesy of ESAB Robotic Welding.)

1.2 COUNTERBALANCED ROBOT STRUCTURES
   (Figure 1.14)

Another important consideration in the mechanical design of a robot is counterbalancing. The weight of the tools held by a manipulator (such as power screwdrivers, riveters, and welding torches) places a considerable load on the drive system and introduces errors in the motor drive control system. These errors could generate sufficient servo offset to affect substantially the stability of the mechanical drive system. A counterbalance is desirable to compensate automatically for various tool weights.

   A common method of providing counterbalance forces in manipulators is the use of counterweights. Counterweights help in the manual teaching of a robot, but their use introduces substantial

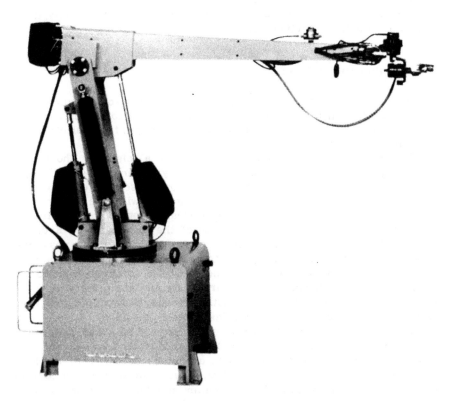

FIGURE 1.14 Robot with reduced weight of the moving parts and a balanced mechanism. (Courtesy of Tokico America, Inc.)

additional mass into the arm, which requires more powerful actuators. Very often the counterweights complicate the design of the arm structure needed to accommodate the added loading.

A spring may also be used to obtain counterweight effects. However, the use of springs is disadvantageous in applications requiring a constant counterbalancing force. Another method of counterweighting is to use an air pressure counterbalance system in which an air-driven piston operates in the vertical direction as the manipulator is moved up and down. The weights of different tools carried by the robot may also be programmed into computer memory so that the air pressure regulator adjusts the counterbalancing force according to the tool used.

The configurations of some robots make it necessary to use a motor to lift both the robot structure and the object held in the hand. An example is the prosthetic robot type. Conventional motors are difficult to use for this type of application. In order to reduce the entire load applied to the motor driving a vertically moving arm of a robot, a pneumatic cylinder may be used to apply a constant adjustable vertical force.

1.3   TYPES OF ACTUATORS

There are three methods for moving the links of a robot, which provide another system of robot classification. Thus robots may be classified by the type of actuator drives at the joints as electric, hydraulic, or pneumatic.

Electrically driven actuators tend to have a relatively short useful life span, primarily because of the heavy loading to which they are subjected. There are inherent time lags present in electric drive motors and associated circuitry which, together with the heavy moving parts, inhibit fast motion. In electrically powered robots the cable must be wired through the joints of the manipulator. Even with sufficient slack in the electrical cable, there is still a possibility of accidents, especially during the teaching and programming stages. Because of such difficulties it may be desirable to have electrical power available in direct form and carried by a means other than cables at the revolute joint. Magnetic or capacitive coupling at the joints used for this purpose provides greater reliability.

Hydraulic actuators are effective for large robots and heavy loads. Because of their large size, lower speeds, and low degree of positional accuracy, hydraulic robots cannot be used effectively in precision manufacturing and assembly. Hydraulic drives are commonly used because they provide high power and force. Pneumatically driven robots are generally smaller and lighter than hydraulically driven robots. However, because of poor position and speed control, the pneumatic drive is less common.

The characteristic features of these three types of robot actuators may be summarized as follows.

Pneumatic actuators:

1. Used in inexpensive manipulators with low load-carrying capacity.
2. Usually require mechanical stops for accurate positioning in non-servo control.
3. Single- and double-acting piston or rotary vane types are commonly used.
4. Inherently lightweight at moderate operating pressures.
5. Limited by low efficiency at reduced loads and low stiffness due to compressibility.

Hydraulic actuators:

1. Linear piston or rotary vane types.
2. High power-to-weight ratio suitable for high loads at moderate speeds.
3. Easily controllable due to high stiffness of hydraulic system.
4. Require high energy storage systems such as pumps and accumulators.
5. Susceptible to leakage, which may reduce efficiency.
6. Require filtration.
7. Air entrapment and cavitation may cause difficulties.
8. Servo valves controlling the fluid flow must be placed close to the actuators to increase the stiffness of the control system.
9. Suitable for harsh environments caused by dust, dirt, and moisture.

Electric motors:

1. Provide a variety of power systems.
2. Easier to control.
3. May not be as responsive as hydraulic systems and stiffer than pneumatic systems with low pressure.
4. Normally require speed reducers, which introduce inefficiency and error in operation.
5. Variety of types such as permanent magnet DC motors, printed-circuit motors, stepping motors, and direct drives.

One of the major factors affecting the characteristics of hydrualic actuators is the mode of hydraulic control. The categories of hydraulic control are pressure control, motion control, flow control, locking control, pressure limiting control, and multiple arm control.

The three basic components used to control pressure are system relief valves, sequence valves, and pressure-reducing valves. Motion controls are used to prevent load drift or overrunning. A simple counterbalance valving can be used for motion control of a robot with relatively fixed patterns of movement. Multiaxis movement requires pilot-assisted counterbalanced valves. Flow control is provided by a flow divider, restrictive flow regulators, or combination flow regulators. A flow divider is typically used to maintain constant flow to two actuators. Restrictive flow regulators maintain constant regulated flow even when the load pressure or pump flow to the valve changes. Combination flow regulators act as both restrictive and bypass valves and limit flow to sections of the circuit for accuracy and safety.

Locking controls are used in hydraulic circuits and have pilot check valves to prevent reverse flow and two-way solenoid valves for instantaneous locking with any loss of power. Pressure limiting controls differ from standard pressure controls because they provide overload or maximum-pressure unloading control. Load-sensing compensators also maximize circuit efficiency; they provide sensing of various segments of the circuit and signal the load required for a specific task.

Hydraulic sequencing valves can be used to supplement typical electronic controls for control of the order in which tasks are performed to prevent conflict. Hydraulic sensing permits the coordinated control of two arms working in concert on a task.

A comparison of the features of hydraulic, pneumatic, and electric controls is shown in Table 1.1.

TABLE 1.1   Robot Controls with Hydraulic, Pneumatic, and Electric Systems

Hydraulic	Pneumatic	Electric
Pressure control	Pressure regulators	Current limiters
Motion control	Pneumatic motion controls	Mechanical motion controllers
Flow control	Flow restricters	Current limiters
Locking control	Pneumatic locking controls	Mechanical brake or self locking worm gear

1.4   DRIVE TRANSMISSIONS

In common practice, the drive motors of robots are mounted on the
movable links of the manipulator arm. This arrangement, although
very convenient, complicates the motion dynamics of the arm by the
additional masses. To overcome this disadvantage, all the drives
may be mounted at the base of the manipulator. The motion is then
transmitted to the joints by an appropriate kinematic connecting system such as shafts, gears, chains, or wire cable. This enables a
more lightweight and maneuverable design of the arm. The extended
kinematic transmission devices from the base to the joints may, however, introduce errors in positioning the arm because of the play in
the transmissions and flexibility of the driving system. Chains leading from the individual drives to the links of the manipulator arm
may be provided with a tensioning mechanism to improve the power
transmission. Several mechanical features of industrial robots are
illustrated in Figures 1.15 through 1.21.

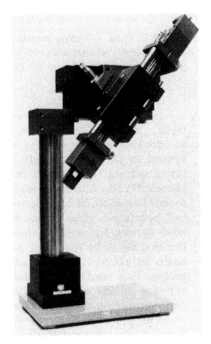

FIGURE 1.15   Modular robots. (Courtesy of Barrington Automation
Ltd.)

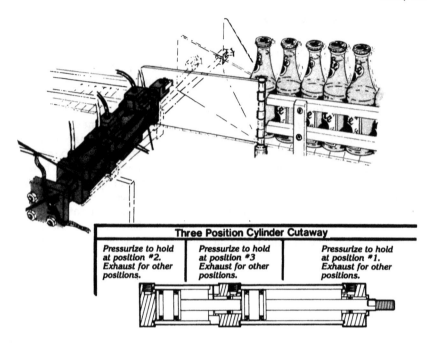

FIGURE 1.16   Three-position linear actuator used to open and close a gate which directs bottles into one of three areas for final packaging.   (Courtesy of PHD, Inc.)

In a typical industrial robot, each chain drive system for the torso, shoulder, and elbow axis includes a motor stage, four stages of sprocket ratio increase, and a final joint stage.   The chain drive system in the wrist module contains the separate stages for the pitch and roll axes, each one having a motor stage, three stages of sprocket ratio increase, and a joint stage.   The chain sprocket sub-assemblies include sprockets, shafts, and end supports that house ball bearings.   The subassemblies are retained by two hexagonal-head screws in each bearing housing.   One screw on each side of the module is in a slotted hole of the module structure to permit chain adjustment.   Chain adjustment is accomplished by adjusting screws that contact the bearing housing, one on each side.   Different-size chains of various lengths are used in each module.   Spring clip connecting links or offset links are used for chain installation.

The placement of hydraulic and electric drive components in the base of a robot contributes to a reduced arm weight.   In such a case, arm and wrist motions may be transmitted through push rod assemblies similar to the control linkages in aircraft.   This type of arrangement eliminates flexing of hydraulic and electric lines at the arm and also allows full arm deflections with minimum movement of hydraulic cylinders.

Common electromechanical drives such as harmonic drives used in commercial robots need gearing to transmit the torques that move the joint axes. Factors like limited mechanical stiffness and torque fluctuation adversely affect the performance of a robot. Backlash problems in gearing can often be alleviated by preloading. If a large preload is used, however, friction and quick wear may result.

Direct-drive methods offer an alternative that may be used to overcome transmission problems in robot drives. The direct-drive method is a straightforward way to achieve high-quality dynamic performance of manipulators. It departs radically from conventional arm drive mechanisms consisting of gears, chains, and lead screws. In a direct drive, the joint axes are directly coupled to rotors of high-torque electric motors. Because there is no transmission mechanism between the motors and their loads, the direct-drive system has excellent features such as no backlash, small friction, and high

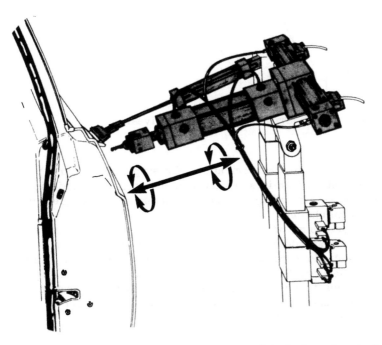

FIGURE 1.17 Multimotion actuator used to lock and unlock a car door for an endurance test. (Courtesy of PHD, Inc.)

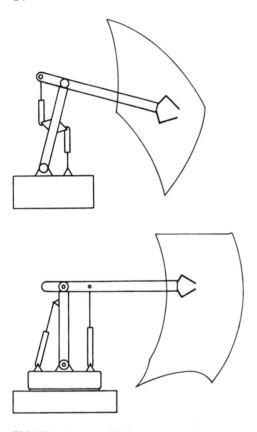

FIGURE 1.18   Robot arms with hydraulic drives.

stiffness. This permits the derivation of a simple, clear, and precise dynamic model which is advantageous for accurate positioning
control and for compensating interactive and nonlinear torques in
high-speed manipulation.

The simpler design of direct-drive motors without any gears,
belts, or other transmissions results in more reliability and less
maintenance and allows robots to be made more compact and of lighter weight. The brushless motors produce no sparks or brush particles. This means that much more use could be made of electric
robots in such tasks as spray painting, because the danger of ex-
plosion becomes minimized.

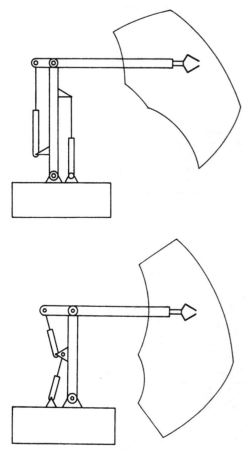

FIGURE 1.18   (Continued)

A distinguishing feature found in some direct-drive robots is the parallel low-friction linkage. Such a linkage balances the loading on all of the joints and therefore reduces the size of the motors. Another advantage of the parallelogram linkage is that it provides a compliance in the end-of-arm tooling. The direct-drive method has the promise of being the answer to high-speed assembly.

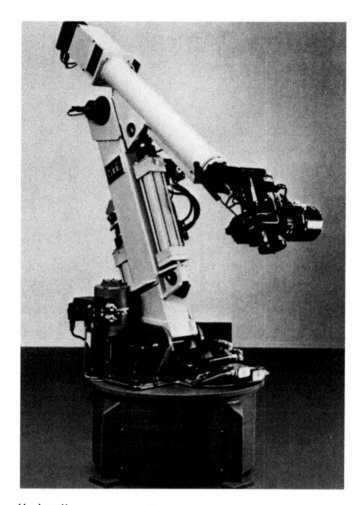

Hydraulic power supply:
    Flow: 64 l/min (17 gpm)
    Operating Pressure: 10.3 MPa (1500 psi)
    Filtration: 3 μm Absolute

Utilities, environmental:
    Electrical: 460 V, 3-phase, 20 hp, 3 kVA
    Cooling water: 11 l/min (2.9 gpm) at 18°C (65°F)
                         (air cooling optional)
    Environment: 0–43°C (32–110°F)

Weight:
    1300 kg (3000 lbs)

FIGURE 1.19   Hydraulically driven robot and its installation requirements.   (Courtesy of MTS Systems Corporation.)

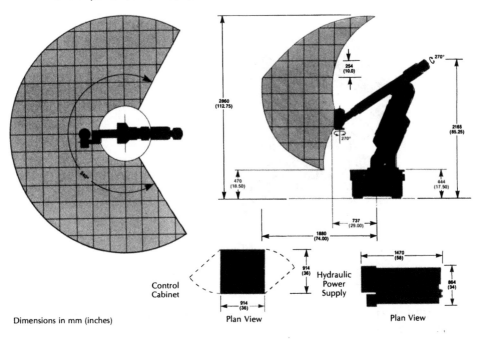

Dimensions in mm (inches)

Control Cabinet

Plan View

Hydraulic Power Supply

Plan View

Specifications
Robot Mechanical System

Degrees of Freedom: 4, 5 or 6 as specified.

Capacity: Load — 100 kg (220 lbs.) at any point within working envelope.
Torque — 450 N-m (4,000 in.-lbs.) wrist torque with 127 mm (5 in.) tool plate offset.

Repeatability: ± 0.13 mm (± 0.005 in.) carrying any weight up to rated load.

Stiffness: 700 N/mm (4,000 lb./in.) at end of arm, with arm fully extended.

Axes Capabilities:

		Speed (degrees per second)
Sweep	240°	0 - 120
Shoulder	55°	0 - 23
Elbow	90°	0 - 35
Elbow Roll	270°	0 - 180
Wrist Roll	270°	0 - 180
Wrist Pitch	270°	0 - 180

(Example max. linear speed 3.8 m/sec, 150 in/sec at 75 in. radius.)

FIGURE 1.20 Geometry and mechanical specifications of a robot. (Courtesy of MTS Systems Corporation.)

FIGURE 1.21   Robot arm with chain transmission.   (Courtesy of
Rhino Robots, Inc.)

1.5   FEATURES OF CYBOTECH ROBOTS*
   (Figure 1.22 a–d)

The V15 is an electric robot. Its compact design and efficient power
unit enable the system to make maximum use of available space. Low
maintenance, high productivity, and outstanding performance are
primary considerations in the manufacture of the V15. It is built
for applications with tool and part loads up to 15 kg (33 lbs).
Available in either five- or six-axis configurations, the V15 is ideal
for welding, parts handling, cutting, drilling, and other operations
demanding precision and reliability. The robot can manuever a 15-
kg load at up to 68 in./sec with a repeatability of ±0.1 mm (0.004
in.).
      The V80 is a versatile, precise industrial robot built for appli-
cations with tool and part loads of up to 80 kg (175 lbs). Its mechanical configuration makes it ideally suited for applications where
horizontal obstacles must be avoided or access to the workpiece is
best from the top. Among the best applications for the V80 are

---

*Reprinted by permission of Cybotech Corporation.

(a)

(b)

FIGURE 1.22   Cybotech robots having different configurations.
(Courtesy of Cybotech Corporation.)

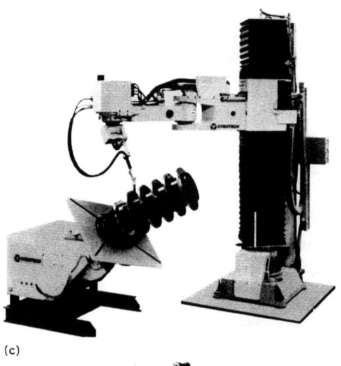

(c)

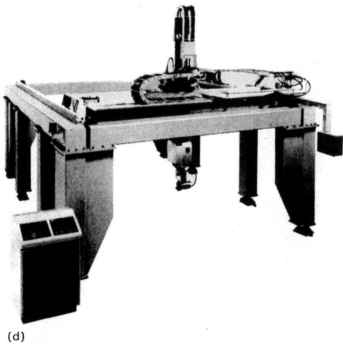

(d)

FIGURE 1.22   (Continued)

parts handling, loading, and palletization tasks. The primary applications for the V80 include resistance welding, arc welding, cutting
operations with various tools, and heavy handling and loading tasks.
The V80 design affords maximum productivity. With an 80-kg load,
repeatability is better than ±0.2 mm (0.008 in.).

The G80 is built for applications with tool and part loads of up
to 80 k (175 lbs). The gantry configuration is ideally suited for
applications where overhead mounting is advantageous, such as trans-
er line operation. It is available in either electric or hydraulic versions. The G80 can be equipped to perform tasks such as arc welding, spot welding, drilling, and cutting with torch, plasma arc, or
water jet. The G80 design affords maximum productivity. With an
80-kg load, repeatability is better than ±0.2 mm (0.008 in.). Multiple G80 robots can be suspended from the same gantry framework.
This is appropriate for assembly line applications where similar or
identical operations are required on both sides of a workpiece.

The H80 is a versatile, precise industrial robot built for applications with tool and part loads of up to 80 k (175 lbs). Its horizontal reach and vertical travel make it well suited for assembly line
applications requiring work on large parts, such as spot welding of
auto or truck bodies, and for heavy-duty continuous arc welding.
The primary applications for the H80 include arc welding, cutting
operations with various tools, and heavy handling and loading tasks.
It is available in either electric or hydraulic versions. The H80 design affords maximum productivity. With an 80-kg load, repeatability is better than ±0.2 mm (0.008 in.).

1.6   FEATURES OF THE IRI M50 ROBOT*

The IRI M50 robot is a self-contained lightweight five-axis unit that
employs air motors for its motive force in conjunction with chain
drives to obtain its operating strength and speed (Figure 1-23). It
has been designed and constructed in modules. High-strength aluminum alloy structures are used to provide the required tensile
strength. Four functional modules (torso, shoulder, elbow, and
wrist) contain the air motors, valve assemblies, encoders, brakes,
sprocket assemblies, chains, wiring, and plumbing for the associated
axis. The majority of the parts and components are interchangeable
among these modules. An equipment module containing the pneumatic
pressure control assembly and the pneumatic connection is located
adjacent to the spindle box assembly. The spindle box assembly
supports the spindle shaft and ties the torso, shoulder, and equipment modules together. (See Tables 1.2 through 1.4 for
specifications.)

---

*Reprinted by permission of International Robomation/Intelligence.

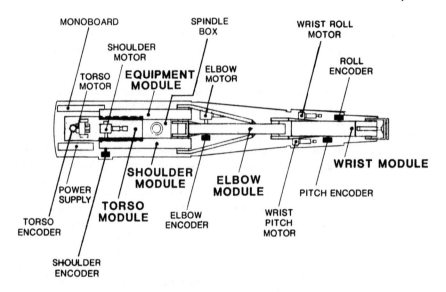

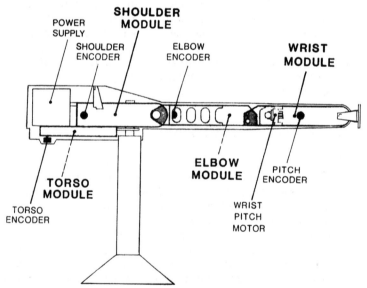

FIGURE 1.23   Robot module identification.   (Courtesy of International Robomation/Intelligence.)

TABLE 1.2   Robot Specifications

Item	Specification
Robot arm	
Axis	Five axis, standard
Clearance required	Sphere, 80-in. radius
Weight	400 lbs (without optional base)
Drive	Digital air servo motor
Maximum load	50 lbs (including end effector)
Reach	Minimum 20 in./maximum 80 in.
Repeatability	±0.040 in. (±1 mm)
Motion range and maximum speed	
Torso	360° 1.5 revolutions, 60° per second
Shoulder	180° ± 90°, 30° per second
Elbow	230° ± 90° − 140°, 45° per second
Wrist—pitch	240° ± 120°, 120° per second
Wrist—roll	360° ± 2 revolutions, 120° per second
Control	
Microcomputer	Motorola 68000 w/ 32KB memory
(torso mounted)	Five independent axis computers dedicated safety computers
Input/output ports	Eight input/eight output (optional 16/16)
Communications	Two independent RS232C connections
Programming	
Language	Robot command language (RCL)
Modes	On line; teach—playback
Environment	
Ambient temperature	40°C (104°F) max.
Relative humidity	90% max--noncondensing
Power inputs	
Electrical	115 VAC, 50−60 Hz 5A
Air	100 psi, 20 to 130 scfm, 1-in. supply line
Options	
Base	48 in. high, 24 × 24 in. base
Universal end effector	Two-finger compliant hand
Memory	Up to 128k bytes
Move pendant	
System terminal	Keyboard, hard-copy printer
Load terminal	Microcassette
Utility I/O module	Eight digital in, eight digital out
Base installation kit	
Pneumatic installation kit	FLR—filter, lubricator, regulator
Backup battery	Supply to RAM memory

Source: Courtesy of International Robomation/Intelligence.

TABLE 1.3 Axis Numbering

Axis	Description
1	Torso (body sweep)—rotates the robot around the vertical axis (spindle) with ± continuous range at a maximum speed of 60° per second.
2	Shoulder—rotates the upper arm up or down around the centerline of the shoulder shaft, with a motion range of 180° (up 90°, down 90°) at a maximum speed of 30° per second.
3	Elbow—rotates the forearm up or down around the centerline of the elbow shaft, with a motion range of 230° (up 90°, down 140°) at a maximum speed of 60° per second.
4	Pitch (wrist)—rotates the wrist faceplate up or down around the centerline of the wrist shaft with a motion range of 240° (up 120°, down 120°) at a maximum speed of 120° per second.
5	Roll (wrist)—rotates the wrist plate about the longitudinal centerline of the forearm with ± continuous range at a maximum speed of 120° per second.

Source: Courtesy of International Robomation/Intelligence.

TABLE 1.4 Module Description

Module	Functional description
Torso	Located on the underside of the spindle box assembly; contains the chain drive system for rotating the robot around its vertical axis.
Equipment	Located on the left side of the spindle box assembly; contains the pressure control assembly, pneumatic supply connection, electronic and computer control mounting, and radial bearing for the left side of the shoulder shaft.
Shoulder	Located on the right side of the spindle box assembly; contains the chain drive system for rotating the shoulder shaft.
Elbow (upper arm)	Attached at the shoulder shaft; contains the chain drive system for rotating the elbow shaft.
Wrist (forearm)	Attached to the elbow shaft; contains the chain drive systems for the wrist shaft for pitch and roll axes.
Spindle box assembly	Contains the spindle shaft assembly and serves as structural support for torso, shoulder, and equipment modules.

Source: Courtesy of International Robomation/Intelligence.

Six separate molded fiberglass skin covers completely enclose the IRI M50 robot. The covers are in mating halves, with one pair for the torso/shoulder modules and one each for the upper arm and forearm.

Torso Module (Axis 1)

The torso module is bolted at its forward end to the spindle box assembly. Its forward sides are bolted on one side to the equipment module and on the opposite side to the shoulder module. The torso module contains the chain stages for rotating the robot about its vertical axis. Holes on the sides, top, and bottom of the torso module provide access for inspection and maintenance, such as chain adjustments. The air motor/valve assembly is mounted on top of the module at the aft end. The dual brakes are on the top and bottom of the sprocket at stage 1, while the encoder is installed at the motor stage.

Shoulder Module (Axis 2)

The shoulder module is on the right side of the spindle box assembly and the forward right side of the torso module. It contains the chain stages for rotating the shoulder shaft subassembly. The right side of the shoulder shaft assembly is supported in the module by thrust washers, two thrust bearings, and two radial needle bearings. Access holes are provided for chain adjustment and maintenance. The dual brakes are installed at stage 1, while the encoder is at the motor stage.

Elbow Module (Axis 3)

The elbow module hinges on the shoulder shaft and is supported by bathtub fittings bolted to the shoulder shaft subassembly. Side braces rotate on the ends of the shoulder shaft and are attached to the sides of the module for additional support. The module contains the chain stages for rotating the elbow shaft inside the two thrust bearings and two radial bearings installed in the forward end of the module. Access holes are provided for chain adjustment and maintenance. The dual brakes are installed at stage 1, and the encoder is installed at the motor stage.

Wrist Module (Axes 4 and 5)

The wrist module is attached to the elbow shaft by means of a bolted clamp. It contains the chain stages for pitch and roll axes. Access to the chains for adjustment and maintenance is through

acces holes on the sides and top of the module. The air motor/
valve assembly and encoder for the pitch are on the right side,
while for the roll they are on the left. Brakes are installed at
stage 1 for each axis and the encoders are at stage 2.

The chain drive for the pitch axis rotates the wrist shaft,
which is installed inside needle bearings and thrust bearings. The
chain drive for the roll axis rotates the hand by a bevel gear arrangement. Access to the wrist shaft assembly is by removing the
front cover on the module. A nylon filler prevents nominal foreign
objects from entering the wrist shaft assembly.

Equipment Module

The equipment module is installed on the left side of the spindle
box assembly and the forward left side of the torso module. It contains the pressure control assembly and two needle bearing supports
for the left side of the shoulder shaft. Access to the pressure control assembly components is by a removable top cover. This module
also contains the monoboard assembly.

Spindle Box Assembly

The spindle box assembly is bolted to the equipment, shoulder, and
torso modules. It rotates around the spindle shaft, which is se-
cured to the robot mounting base. A needle bearing and a thrust
bearing on the spindle at the top and bottom of the box assembly
carry the torso axis loads of the IRI M50. The thrust bearing on
the bottom of the spindle rests on the torso joint sprocket, which
is welded to the spindle shaft. The spindle shaft is retained in the
box by a lock washer and a nut that is tightened against the top
thrust bearing.

Pneumatic System

Air is furnished to the IRI M50 robot at the main manifold in the
equipment module and is distributed to the five axes through separate solenoid-operated pressure control valve assemblies and volume
chamber/accumulators. The air from each volume chamber/accumula-
tor is directed to the motor/valve assembly for the associated axis.
Air from the main manifold is also directed to each axis brake
through individual on/off electrical valves.

A pressure regulator is connected to the main manifold and
regulates air directed to the pilot pressure manifold. The pilot
pressure air is regulated at 80 psi and is supplied to the motor/
valve assemblies, furnishing a controlled air pressure for each axis.
The spool position in the valve assemblies determines the direction

of air flow to the air motors. The spool is positioned as a result of solenoid valve operation. Energizing a solenoid opens a poppet valve, which directs pilot pressure to move a piston inside the valve end, causing pressure to be exerted on the end of the spool valve and moving it to the required position. The air motor direction is reversed by operation of the solenoid on the opposite end of the valve assembly.

Pressure Control

Five pressure control valves are installed in the equipment module, one for each axis. Each control valve assembly contains a solenoid mounted on a valve end and a two-position valve that is operated by pilot pressure when the solenoid is operated. A dithering technique (pulsed air) is used to control the air pressure in the accumulators so that the supply pressure for each motor is constant. The air pressure is modified for different accelerations or to maintain a velocity.

There are also five volume chamber/accumulators installed in the equipment module, one for each axis. Each volume chamber/accumulator maintains the required air pressure for one axis. The pressure is controlled by the pressure control valve assembly on which each volume chamber/accumulator is mounted. In addition to serving as a pressure accumulator, the volume chamber/accumulator dampens the air pulses to the air motor. A connection is provided on each chamber for connecting to the pressure transducers.

Air Servo

An air motor is furnished for each axis. The motor for the elbow and shoulder axis delivers up to 1-1/2 hp, while the one for the torso, pitch, and roll axes delivers up to 1/2 hp. Each motor contains eight vanes and is a reversible type. The vanes are self-sealing and self-adjusting. Exhaust air cools the air motor as it turns, enabling it to be used at temperatures up to 250°F. Lubrication is required by means of an air line lubricator in the supply line. Each air motor attaches to its motor stage sprocket assembly by a shaft extending into the sprocket assembly and pinned to it.

The air motor valves are spool type, three-position, in-line-spring centered, contain five ports, and are operated by solenoid-operated pilot-poppet valves. The ports are numbered for each of identification: 1, pressure; 2, air motor; 3, exhaust; 4, air motor; and 5, exhaust. The identification plate indicates the ports that are connected when each solenoid is activated. Mufflers are installed on the exhaust ports for noise dampening. A manifold mates each air motor and valve assembly.

Brakes

Two brake assemblies act as a caliper on the stage 1 chain sprocket
for each axis. All brakes are pneumatically operated except the
backup brake for the pitch axis, which is stationary. An electrical
control valve is furnished for each axis and directs the air pressure
to the brake(s). The pneumatically operated brakes have piston
assemblies retained in aluminum alloy housings and contain O-ring
seals. All brakes are made with wire asbestos brake material and
are bonded to an aluminum alloy housing. Adjustment is provided
for the stationary backup brake used on the pitch axis.

1.7   MECHANICAL FEATURES OF SPINE
         ROBOTS*

A Spine robot comprises two main parts: an under arm with fittings
and an upper arm with intermediate disc and wrist. The base section rests on a base plate, which contains distribution channels for
the hydraulic oil. The base contains the four hydraulic cylinders
for the under arm, the valve block and filter, and position sensors
for the under arm. The arm joint contains the cylinders and the
position sensors for the upper arm. Power sensors are fitted in all
the hydraulic cylinders. See Figures 1.24 through 1.27 for different mechanical features.

The under arm and upper arm comprise a number of discs with
hemispherical contact surfaces—that is, what is known as the cross-
plate and the discs. Four wires are connected to the under arm
and a similar number to the upper arm. The tension corresponds
to a force of several tons per wire pair. The under arm's wires are
attached to each hydraulic cylinder in the lower section, and the
wires in the upper arm are connected to the hydraulic cylinders in
the arm joint. The cylinders are under constant pressure from the
hydraulic system and this pressure gives rise to the actual wire
tension. Sensors attached to the cylinders measure the forces in
the respective wires.

The Spine robot's under and upper arms are constructed of a
number of crossplates and discs. The prestressed lines run through
the crossplates—four in the lower arm and four in the upper arm.
The wires are the robot's "muscles" and the arm has the necessary
rigidity because of the tension. The four wires in each arm form
two pairs. When one wire pair is fixed and the force in the other
pair is increased or decreased, the arm will rotate about *one* axis.
An axis is formed by the wire pair where the change in force takes
place.

---

*Reprinted here by permission of Spine Robotics Corp.

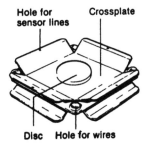

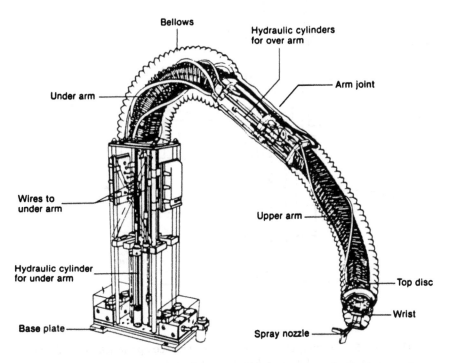

FIGURE 1.24 Mechanical design of Spine robots. (Courtesy of Spine Robotics Corp.)

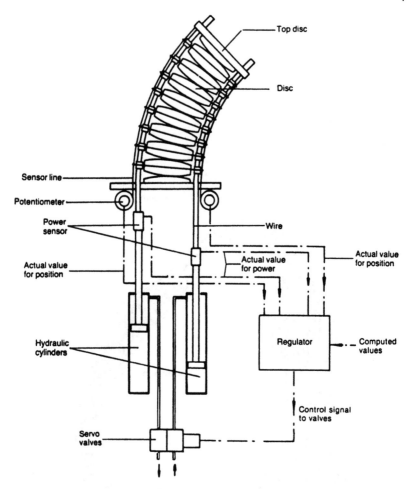

FIGURE 1.25 Arrangement of the control components in a Spine robot. (Courtesy of Spine Robotics Corp.)

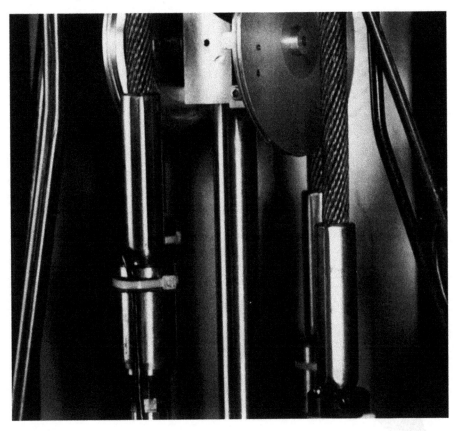

FIGURE 1.26   Mechanical design and construction features of Spine robots. (Courtesy of Spine Robotics Corp.)

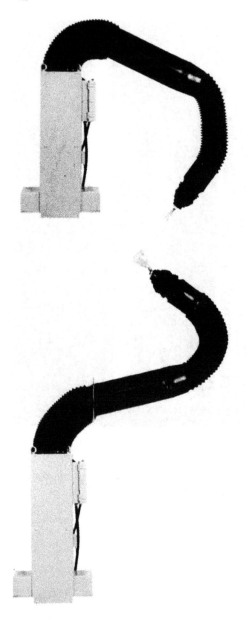

FIGURE 1.27   Mechanical flexibility of a Spine robot. (Courtesy of Spine Robotics Corp.)

FIGURE 1.27 (Continued)

REVIEW EXERCISES

1. Explain what is meant by a robot.
2. Describe the four common types of geometric configurations of
    a robot and give some examples of industrial robots belonging
    to these four common types. Sketch the workspaces of these
    robots in three-dimensional and orthographic projection views.
3. Find two examples of robots that have closed linkage geometric
    arrangements. Discuss the advantages of this type of robot.
4. What are parallelogram linkage robots? Sketch the workspaces
    of a typical robot of this type.
5. Explain the following terms: pendulum robot, gantry robot,
    serpentine robot, and mobile robot.
6. Describe, compare, and contrast the features of hydraulic,
    electric, and pneumatic actuators.
7. Explain briefly the operation of harmonic drives and direct
    drives in actuating robots.
8. Discuss the methods of counterbalancing of robot links.
9. Discuss the detailed mechanical features of two industrial
    robots.
10. Write a short paper summarizing the material in some of the
    selected bibliographic references for this chapter.
11. Use the most recently published manufacturers' brochures to
    classify a number of current industrial robots by their coordi-
    nate system.
12. Study the table of specifications and produce a comparative
    table describing the features of two robots.
13. Study the mechanical components including actuators, trans-
    mission, and braking in four typical robots.

FURTHER READING

Asada, H., "Development of a Direct-Drive Robot and Evaluation of
    Its Control Performance," *Soc. Instrum. Control Eng. Trans.*,
    Vol. 19, No. 1, 1983, pp. 77–84 (in Japanese).
Asada, H., et al., "A Direct Drive Manipulator—Development of a
    High Speed Manipulator," in *Developments in Robotics*, IFS
    Publications, Bedford, UK, 1983, pp. 217–226.
Barber, N. T., "Benefits of Using Brushless Drives on High Per-
    formance Robots," *Proc. 7th Brit. Robot Assoc. Annu. Conf.*,
    Cambridge, UK, May 1984, pp. 95–106.
Bruchmann, K., "Special Motors for Industrial Robots," *Elektr.
    Masch.*, Vol. 62, No. 10, 1983, pp. 275–278 (in German).
Davies, B. L. and Ihnatowicz, E., "Hydraulic Power for Industrial

Robot Manipulators," *Proc. Int. Fluid Power Symp.*, Cambridge, UK, April 1981, pp. 361–372.

Engelberger, J., *Robotics in Practice—Management and Applications of Industrial Robots*, American Management, New York, 1981.

Hammett, G. G., "Low Cost Servoactuator Design," *Proc. 8th Int. Conf. Industrial Robots*, Detroit, June 1984, pp. 4.45–4.54.

Heer, E., "Robots and Manipulators," *Mech. Eng.*, Vol. 103, No. 11, 1981, pp. 42–49.

Heginbotham, W. B., "Present Trends, Applications and Future Prospects for the Use of Industrial Robots," *Proc. Inst. Mech. Eng.*, Vol. 195, 1981, pp. 409–418.

Kinoshita, H., Kake, N. and Yoshitake, H., "AC Servo System for Robots," *Natl. Tech. Rep.*, Vol. 29, No. 4, 1983, pp. 623–629 (in Japanese).

Koren, Y. and Ulsoy, A. G., "DC Motor Drives for Robot Arms," American Society of Material Engineers Paper No. 81-WA/DSC-16, 1981.

Lakota, N. A., "Investigations of Mechanical Transmissions for Industrial Robot Modules," *Sov. Eng. Res.*, Vol. 2, No. 10, 1982, pp. 5–7.

Maezawa, T., Yoshizawa, A., and Iwakabe, E., "Structures and Characteristics of Cables for Robots," *Proc. 31st Int. Wire Cable Symp.*, Nov. 1982, pp. 198–206.

Mathisen, E. and Mathisen, E. S., "Electric Drive Robot," *IBM Tech. Disclosure Bull.*, Vol. 26, 1983, pp. 1457–1460.

Menges, G. and Ermert, W., "Industrial Robots," *Kunstst. Ger. Plast.*, Vol. 71, No. 10, 1981, pp. 54–57.

Obrzut, J. J., "Robotics Extends a Helping Hand," *Iron Age*, Vol. 225, No. 9, 1982, pp. 59–83.

Paroi, A. A., "Calculating the Coordinates at Which to Start the Braking of the Piston on the Pneumatic Drive of an Industrial Robot," *Sov. Eng. Res.*, Vol. 1, No. 7, 1981, pp. 7–10.

Rodgers, J., "Compact Air Cylinders Aid Robot Design," *Power Transm. Des.*, Vol. 25, No. 5, 1983, pp. 33–34.

Schulze, M., Emmerling, F., and Haberland, K., "Electrical Axle Drives for Industrial Robots," *Wisee.-Tech. Inf. VEB Kombinat Automatisierungsaniagenbau*, Vol. 18, No. 5, 1982, pp. 210–213 (in German).

Scully, K. R., "Robotics: An Introduction to Today's Robot and Future Trends," Report No. DTNSRDC/CMLD – 83-20, David W. Taylor Naval Ship Research and Development Center, 1983.

Welburn, R., "Ultra High Torque Motor System for Direct Drive Robotics," *Proc. 8th Int. Conf. Industrial Robots*, Detroit, June 1984, pp. 19.63–19.71.

Worsley, R., "What Is a Robot?," *Metalwork Prod.*, Vol. 127, No. 2, 1983, pp. 61–64, 67.

2
End Effectors

The importance of end effectors stems from the fact that they are the means by which a robot accomplishes its task. Careful thought, time, and engineering must go into the design and selection of the end effector in order to utilize the full capabilities of a robot.

The end effector or end-of-arm tooling is the device that a robot manipulates to perform a given task (Figure 2.1). The term end effector covers a wide range of different tools and devices that can be classified in two main categories: (1) grippers and (2) tools. Grippers are the end effectors that actually grip a part for transfer operations in the work envelope of the robot (Figures 2.2 through 2.4). They are normally used for point-to-point tasks such as loading and unloading of machines and palletizing. Tools are the devices which actually perform the task; all the robot does is to position these tools in the working range. Examples of such tools are gluing guns, drills, routers, welding torches, arc welding guns, grinders, spray guns, deburring tools, and automatic screw-drivers. The robot trajectory is important when tool-type end effectors are used. Typically the robot must be programmed to manipulate the tool in a well-defined task path. The following sections examine several aspects of the design and use of grippers.

2.1 DESCRIPTION OF GRIPPERS

Grippers can be classified in four basic categories:

1. Mechanical clamping
2. Vacuum suction

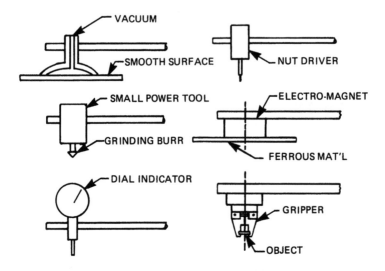

FIGURE 2.1   Typical examples of end effectors.   (Courtesy of Mack Corporation.)

3.   Magnetic attraction
4.   Universal grippers

Mechanical grippers may have two or three fingers (Figure 2.5), although the majority are of the two-finger type.   Triple fingers are suitable for grasping cylindrical and spherical shapes.   Clam fingers pick up objects which are otherwise difficult to grasp.   Shovel-type mechanical fingers work as a digging bucket.   Mechanical fingers may hold an object by grasping from the outside or inside of the object.

Kinematically, mechanical fingers can be differentiated into the following classes:

1.   Linkage type
2.   Gear and rack type
3.   Cam type
4.   Screw type
5.   Rope and pulley type

Vacuum cups or pads made of rubberlike materials are ideal for nonferrous objects with flat and smooth surfaces, curved surfaces, and textured surfaces, including glass, large sheet metal, and paper products.   The use of single or multiple vacuum cups is dictated by

FIGURE 2.2   Mechanical gripper.   (Courtesy of Mack Corporation.)

the size and weight of the object to be handled by the robot. Multiple cups maintain the orientation of large parts. If the workpiece is small and light enough, vacuum grippers without a vacuum pump are adequate. However, larger and heavier objects require a vacuum pump or a venturi vacuum generator to operate the vacuum cup. Delicate objects, such as sunglass lens, are better handled by vacuum cups.

Magnetic grippers use a magnetic head constructed with a ferromagnetic core and conduction coils to attract ferrous materials. Although magnetic grippers have greater efficiency and reliability, the residual magnetism in an object handled may pose a problem.

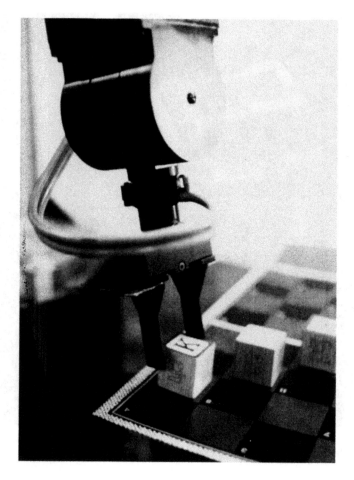

FIGURE 2.3   Gripper with two parallel fingers for holding an object with flat sides.

Another disadvantage is that magnetic grippers can pick up more than one part from a bin of parts.

In the class of universal grippers can be found inflatable fingers, soft fingers, and fingers made from moldable materials. Inflatable grippers are used to pick up fragile and irregular objects with an evenly distributed load. The inflatable bags may be made of a thin rubber which holds a loose medium such as metallic or plastic spherical particles. Inflatable bladder-type grippers are expanded by pressurization after the gripper has been brought into

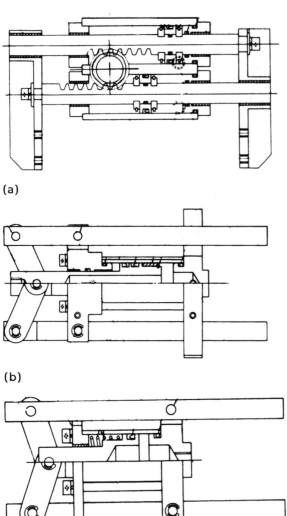

(a)

(b)

(c)

FIGURE 2.4   Mechanical design of grippers.   (Courtesy of
JERGENS/NKE.)

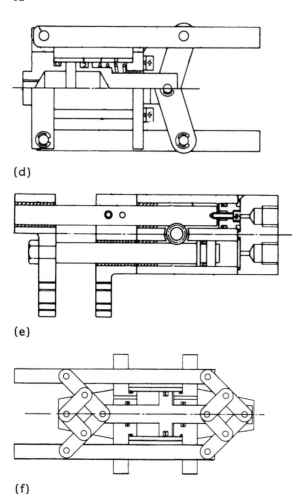

(d)

(e)

(f)

FIGURE 2.4   (Continued)

position. If the expansion is inward the gripper surrounds part of
the object. Soft fingers consist of multilinks and a series of pulleys
actuated by a pair of wires. It is easy for a soft gripper to conform with uniform pressure to the periphery of an object having a
concave or convex surface.

The nature of the gripping surface is an essential factor for ensuring good robot performance. A material very commonly used for
gripping pads is a polyurethane bonded to steel. The high coefficient of friction of the polyurethane produces high gripping efficiency.

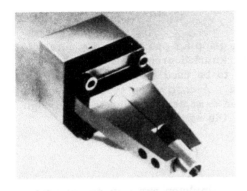

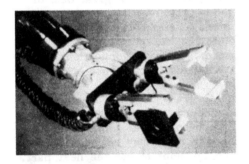

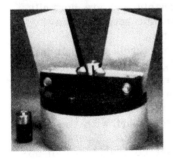

FIGURE 2.5 Double-acting two- and three-jaw styles of grippers.
(Courtesy of Compact Air Products, Inc., Westminster, S. Carolina.)

In an application which calls for a gripping pad to actually bite into a surface, hardened surfaces with knurled projections should be used. A groove or slot may have to be used to ensure positive grip if the friction grip is not adequate. A captivation-type grip should be used for picking up an egg, for example, in order to avoid high clamping forces which may damage the egg.

2.2   DUAL GRIPPERS

Installation of dual grippers on a manipulator arm may be required in the following situations:

1.   Dual workstations
2.   Part size variation caused by the machining process
3.   Processing of two different-size parts
4.   Machine downtime reduction
5.   Increased production rate requirements

Machine tools equipped with dual workstations increase productivity significantly. Dual grippers maximize the use of such dual workstations. In machining operations, removal of a great deal of material from the green part may reduce the overall size of the part by a large percentage. Wide variations of size between the green part and the finished part may dictate the use of two different grippers in a dual-gripper arrangement. In some applications the parts handled may have sufficient variations that the use of dual grippers is imperative.

A dual gripper may be designed to grasp both the inside diameter and the outside diameter of a workpiece. This allows a robot to unload a machine with one gripper while holding an unprocessed workpiece in the other. In this way the robot may reload the machine without moving back to the pickup point to get the next part. Dual grippers are attached to the wrist assembly and are normally actuated pneumatically or electrically. Hydraulic actuation may, however, be needed in handling heavy objects.

The production rate of a robot with a long cycle time can be increased by using multiple part grippers (Figure 2.6). Downtime occurs when there is no idle workstation for a robot to attend during the machining cycle. In such a situation, a robot with a dual gripper minimizes the machine downtime by removing the finished part with one gripper and quickly loading the green part with the other gripper.

Two limitations of dual grippers that must be considered in applications are weight and size. Because dual grippers increase the end-of-arm payload, a large robot may have to be selected. The

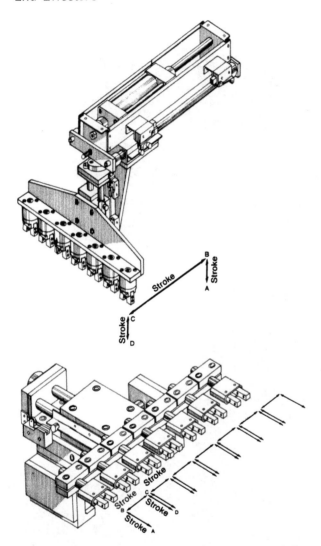

FIGURE 2.6   Multiple and in-line multiple parts transfer with
multiple grippers.   (Courtesy of JERGENS/NKE.)

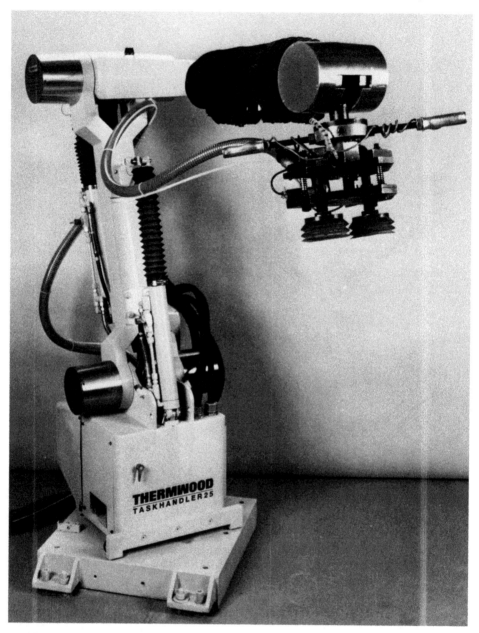

FIGURE 2.7   Double gripper.   (Courtesy of Thermwood
Corporation.)

size of dual grippers also presents clearance problems with tool holders, slides, guides, and other obstructions. This problem may be solved by using different gripper mountings, different robot configurations, and different wrist motions.

Conceptually, two other types of dual grippers can be configured. The first arrangement is really two robots operated by one controller and one power supply. In a typical example involving a forging application, the two-arm robot improves the cycle time. The second approach, which is normally used with cylindrical and Cartesian robots, utilizes grippers on both ends of the robot arm. Advantages of this type of arrangement fall in three categories: cost savings, increased productivity, and tooling simplification. Typical applications requiring this arrangement of grippers can be found in palletizing and stacking tasks using a dedicated dunnage and palletizing with liners or slip sheets between product layers. (See Figures 2.7 through 2.10 for examples.)

FIGURE 2.8   Dual gripper for removing finished spindle and loading new workpiece. (Courtesy of Cincinnati Milacron.)

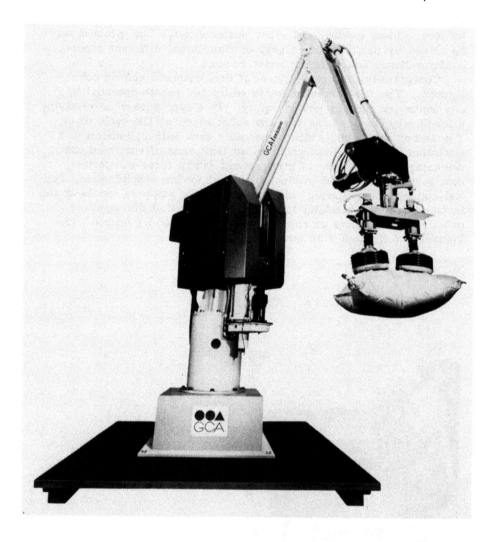

FIGURE 2.9   Dual vacuum grippers for handling cement bags.
(Courtesy of GCA Corporation.)

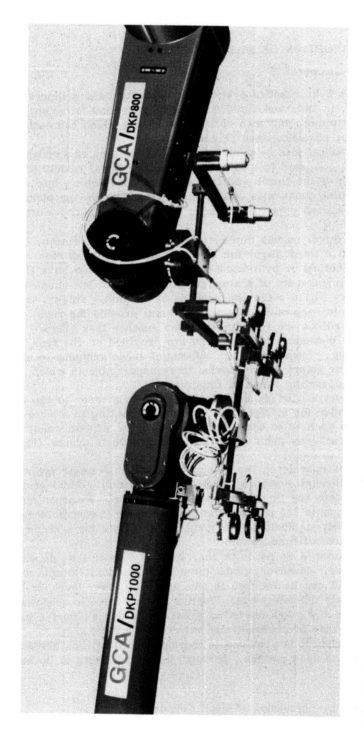

FIGURE 2.10   Multiple gripper.   (Courtesy of GCA Corporation.)

2.3   DESCRIPTION OF SELECTED GRIPPERS

B.A.S.E. Grippers*

In the B.A.S.E. robotics system there are standard grippers in
four sizes, in two- and three-finger configurations for external and
internal gripping, and with soft blank fingers which can easily be
modified for special shapes (Figure 2.11).

The smallest size is about twice the dimensions of a thimble and
the largest compares to a human hand. Two-finger models simulate
the motions of the thumb and index finger for reaching into channels, grasping parts with closely spaced components, or picking and
placing any small object with simple geometry of length, width, and
thickness.

Three-finger models duplicate the motions of the thumb, index
finger, and a third finger for grasping bodies of revolution and objects of spherical or cylindrical shape. Both versions have the
self-centering feature of a standard two- or three-jaw chuck with
repeatability within a few thousandths of an inch. Fingers are interchangeable in each of four model sizes and provide for rapid change-
over when robots are reprogrammed for another task (Figure 2.12).

A pilot diameter and bolt circle are provided on the rear face
for mounting. Simple adaptors will mount these grippers to almost
any linear or rotary motion device to transport objects which are
within the operating range of finger travel.

Transport can be as simple as mounting a gripper to the rod of
an air cylinder for straight-line motion or mounting to a more complex device such as to the "business end" of a six-component manipulator to reach any point in a Cartesian coordinate system (Figure
2.13).

All units operate on the principle of a double-acting cylinder
controlled through a simple four-way valve circuit. Fluid pressure
opens or closes the finger for positive operation. Maximum operating pressure is 150 psi in either hydraulic or pneumatic service;
however, most applications use plant air at 80 psi for a simple and
reliable source of fluid power.

Pinch force at 80 psi is in the range of 5 to 50 lbs, depending
on model size. However, pinch force is proportional to operating
pressure and can be reduced to any nominally lower level for handling extremely delicate parts. Further, pinch force is essentially
constant for any given operating pressure, which allows the gripper
to grasp objects of different size with uniform force.

An adjustment is provided to increase the dimension across the
fingers in the open position, permitting larger objects to be handled

---

*Reprinted by permission of Mack Corporation.

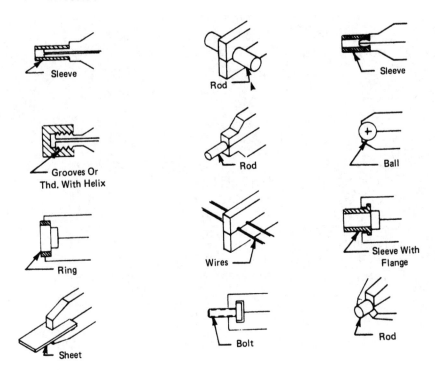

FIGURE 2.11   Gripper designs for various objects.   (Courtesy of
Mack Corporation.)

when it is not necessary for the fingers to touch in the closed
position.

Overall construction is as light as practical for high-speed cycling and rapid transport.   The smallest units weigh 2 ounces and
the largest weigh 9 ounces.   Pistons and adjusting stems are stainless steel; bodies, covers, and fingers are made from high-strength
aluminum alloy.   Seals are standard O-rings.   Barbed fittings for
1/8-in. flexible tubing are included with each unit.

Features of ISI Vacuum Grippers*

Vacuum cups are manufactured in a variety of shapes and sizes.
Most common are oval and round cups that range in size from

---

*Adapted here by courtesy of ISI Manufacturing Inc.

FIGURE 2.12   Various finger arrangements in robot grippers.
(Courtesy of Mack Corporation.)

1 to 5 in. in diameter (round) and 2 to 8 in. (oval). See Figure
2.14 for examples.

Most cups are available in a durometer range of 30 to 60 and
are molded from either neoprene or urethane compound. Look for a
cup that is molded over a steel supporting insert for strength and
offers a reinforced lip to resist cutting or tearing—one with good
memory qualities. Cups flex and distort and must return to their
original shape in every cycle.

One of the most commonly used generators of vacuum is the
venturi. All venturis, regardless of their size or shape, function
in the same manner. The venturi principle is based on Bernoulli's
theory that an increase in velocity produces a decrease in pressure.

Venturis accept standard shop air regulated down to an efficient
22 to 32 psi. Air flowing through the larger orifice creates a vacuum effect through the smaller cross orifice. It is at the smaller orifice that the vacuum cup is attached.

Air consumption is generally 5 to 9 cubic feed per minute.
Properly adjusted vacuum cups will hold 10 lbs per square inch of

cup area; which in turn makes each cup capable of holding 10 to
100 lbs with a 2:1 safety factor.

Multiple cups are each fitted with their own venturi. This will
ensure faster venturi evacuation time and eliminate dropping the
part if the cup is damaged or the air line is severed. There are
two styles of venturis used with vacuum cups

1. Fixed or non adjustable
2. Adjustable

The fixed or block type venturi has a piped tap for attaching
desired vacuum cups. This is assembled by means of a close nipple.
An adjustable system may be desirable when handling a variety of
shapes and sizes. Adjustable venturis, such as the vaclok vacuum
cup and venturi assembly combination, offers a square T-slot

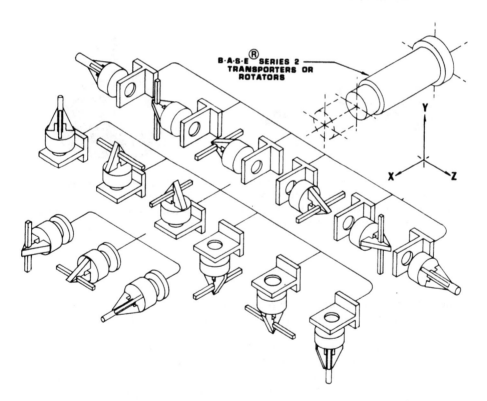

FIGURE 2.13   Mounting options for mechanical attachment of
grippers. (Courtesy of Mack Corporation.)

FIGURE 2.14   Vacuum head assembly permitting easy multiple cup adjustments or quick changeover in handling fragile or finished parts. Nadir silencers reduce venturi exhaust sound below acceptable decibel level. (Courtesy of I.S.I. Manufacturing Inc.)

attaching feature between cup and venturi which quickly detaches for replacement without disturbing the angular positioning of the mount and bracket arm.

The vaclok system is available with an adjustable bracket arm which offers 360° orbital and 30° angular positioning.

Most venturi and bracketry should be of lightweight aluminum but should be of rigid construction to minimize weight at the end of the arm.

Popular and useful options are:

1. Vacuum actuating value: Dual purpose device that fits between the vaclok cup and venturi and offers (a) Part blow off: quick release, upsetting the vacuum seal made by the cup (high speed applications) and (b) Safety seals in vacuum in the event that the vacuum air line is severed.
2. Vacuum heads offer auxiliary – additional lift to lower or raise cups angularly or vertically.

The gripper head is used for the action with links removed. A variety of bases are available.

Considerations or sizing the system include

1. Part overall size
2. Part rigidity
3. Surface irregularities—smoothness, porosity of material (fabric), sharp brakes
4. Part weight
5. Multiple cup placement for balancing the load

Popular vacuum sensing devices commercially available are

1. Vacuum sensing switches (pressure).
2. External proximity or limit switches indicating the presence of the part throughout the handling cycle. These devices may be mounted to the vacuum tooling.

Various I.S.I. gripper types are shown in Figures 2.15 through 2.17.

Robohand Grippers[*]

The air-operated gripper has two moving fingers capable of external or internal surface gripping (Figure 2.18). The body is composed of hard anodized high-strength aluminum, the fingers of hardened 17-4PH stainless steel, and the bottom cup (which houses the dimension sensor) of nonmetallic delrin. A calibrated spring-loaded gap between the body and mounting base provides safety travel for parts misplacement situations, collisions, or force-controlled insertions.

---

[*]Courtesy of Robohand, Inc.

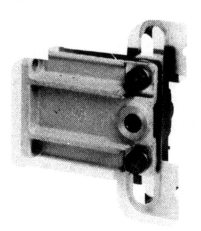

FIGURE 2.15   Parallel motion gripper heads featuring a keyed and tapped machined surface which allows mounting of custom gripping fingers.   (Courtesy of I.S.I. Manufacturing, Inc.)

FIGURE 2.16   Construction features of gripper heads for robots with single, double, or chisel jaw inserts.   (Courtesy of I.S.I. Manufacturing, Inc.)

FIGURE 2.17   Vertical flange and drop-away gripper heads for robots with single or double inserts.   (Courtesy of I.S.I. Manufacturing, Inc.)

The hand is capable of performing part measurement procedures and signaling acceptance or rejection in accordance with the control module's prechosen tolerance.   The set point control of the module permits a varied tolerance range with a high and low set point.   A programmable and adjustable "preset pressure sensor" detects mismating, misaligned, or missing parts.   Both the set point control and present pressure sensor activate relay outputs that can be integrated into system feedback loops.

Through inductive sensing, the RS-25 provides contact closure when a part is not present between the closed fingers.   The inductive proximity switch has no mechanical moving parts and therefore has unlimited life.

2.4   PRINCIPLE OF OPERATION OF PHD GRIPPERS*

In the parallel gripper shown at the top in Figure 2.19 a double-acting piston is attached to a connecting rod.   The cam plate has angled slots machined in it, forming a wedge.   The sliding parallel jaws are attached to the cam plate by pins through the angled slots. As the double-acting piston moves in and out, the sliding jaws are forced to close and open with a parallel motion.

---

*Adapted by courtesy of PHD, Inc.

FIGURE 2.18   Features of Robohand grippers.   (Courtesy of
Robohand, Inc.)

The other parallel gripper shown in Figure 2.19 features a
double-acting piston which is attached to a female socket by a connecting rod.   Each gripper jaw has a lever with a male ball which is
fitted in each side of the female socket.   As air is applied to either
side of the piston, the lever is forced to rotate about a pivot pin.
A second lever or trailing link is attached to the gripper jaw and
body of the gripper.   The result is an arcing motion with the jaw

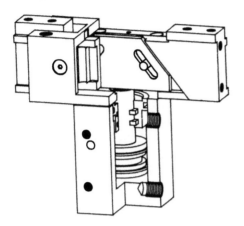

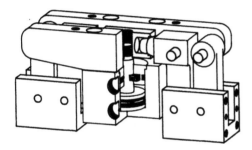

FIGURE 2.19   Operation of parallel grippers.   (Courtesy of PHD, Inc.)

mounting surfaces remaining parallel throughout their gripping movement.

The angular gripper shown at the top of Figure 2.20 has a double-acting piston which is attached to a female socket by a connecting rod. Each gripper jaw has a mating ball fitted in each side of the socket. A pivot pin in the jaw provides the axis of rotation for the angular movement. As air or hydraulic pressure is applied to either end of the piston, the jaw is forced to rotate about the pivot pin by the ball-and-socket assembly, causing opening and closing of the jaw. Another version of an angular gripper is also shown in Figure 2.20.

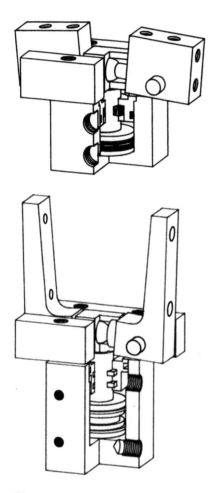

FIGURE 2.20   Operation of angular grippers.   (Courtesy of PHD, Inc.)

2.5   DESIGN FACTORS FOR END EFFECTORS

The basic factors in the design of end effectors for machine loading and material handling applications include the following.

1. Cycle time should be examined in terms of individual cycle time as well as overall system production over a period.
2. Size and mass of parts or tools to be handled should be considered. The size factor may dictate, for example, that the

grippers' fingers move through a range wide enough to accommodate parts with a variety of sizes. The mass factor is important
in the jaw design because of the possibility of parts slipping
out of position during movement.
3. The types of materials to be handled may require, for instance,
nongrasping material-handling hardware. Sheet metal, glass,
and plastic may be handled with suction pumps, while ferrous
parts may be manipulated with permanent magnets or
electromagnets.
4. Some applications may require instrumentation at the end effector to provide pressure feedback, prevent damage to delicate
parts, measure a critical dimension, or sense part presence.

   Characteristics that must be considered in selecting grippers
include parallel or angular jaw movement, type of actuation (pneumatic, hydraulic, or electric), gripper styles, feedback to robots,
size, and force required. Adhesion devices or serrations can be
used to improve the grasping of an object. An encapsulated object
handles better since it is not just the friction which holds it in the
gripper. Sometimes a guide pin is needed to locate the object properly in the gripper. A light bulb, for example, must be handled
by a soft padded gripper. A gripper may be modified for high-
temperature applications by using a ceramic insert heat shield or
drilling for coolant.
   Let us now discuss the selection factors of end effectors for
various tasks (Figure 2.21). Modular end-of-arm components can
be used as the end-of-arm tooling for industrial robots. Some modular grippers can be configured as internal, external, parallel jaw,
or multiple grippers according to the application. Internal gripping
gives flexibility in applications where gripping the outside of the
part is not practical. Parallel grippers offer the advantage of parallel jaw motion, which permits the unit to grip parts of varying sizes
with consistent area and force. Parallel gripping action avoids fixturing and tooling needs for many special applications. Grippers
used in series can handle loads with long axial lengths, which might
twist or escape from only one jaw. (See Figures 2.22 through 2.26).

Application Case Studies

In palletizing rubber bales into shipping containers, a large vacuum
cap may be used for bale pickup. A special puncturing device within the plastic film wrapper helps to prevent the vacuum from rup-
turing the film. In the manufacture of tapered plugs two gripping
hands mounted 180° apart on a channel structure are used. Two
different adjustable finger arrangements provide handling capabilities
for a wide range of plug valve sizes. In the production of forged

diesel engine crankshafts, a robot can be used to transfer the hot
crankshaft from a forging press into a throw twister. A two-forked
hand with pneumatic toggle locking fingers is ideal for such applications. The fork fingers could slide along their support to grasp a
variety of crankshafts at the main bearing positions.

In transferring differential case castings, a hand with a set of
jaws with carbide-tipped inserts for part gripping may be used. In
a cam-operated parallel motion gripper, the piston push stroke opens
the jaws. The pull stroke permits the jaws to close by the action
of extension springs. This gripper is used in feeding billets
through a two-cavity die forging press in the production of automobile differential side gears. In another hand design the clamping
action of the fingerlike paddles is a parallel constant-force motion
generated by two linear cams.

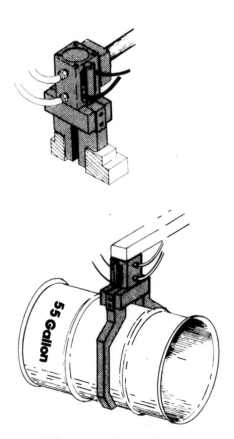

FIGURE 2.21   Features of gripper design for various applications.
(Courtesy of PHD, Inc.)

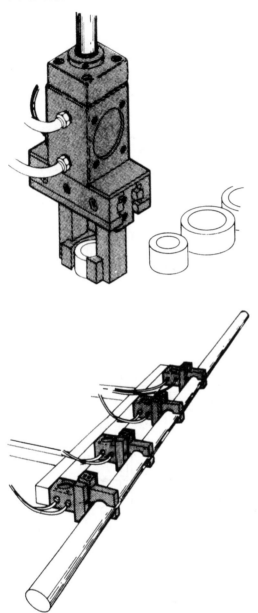

FIGURE 2.21   (Continued)

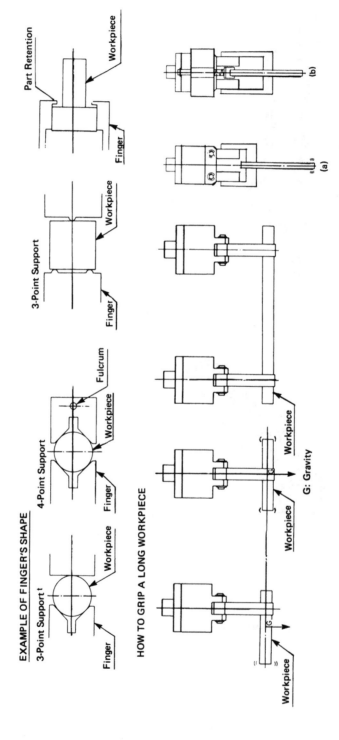

FIGURE 2.22 Design considerations for grippers. (Courtesy of JERGENS/NKE.)

Example Using Center Push Ejector

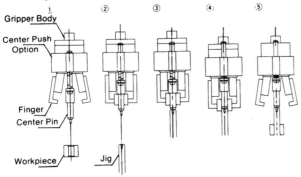

FIGURE 2.23   Gripper with center push ejector.   (Courtesy of
JERGENS /NKE.)

Chain link fabrication and assembly require a robot with special
fingers. The gripper in such an application is a set of fingers with
flat carbide-tipped grippers to hold the chain links.

A special hand for processing gear blanks has two sets of grippers, the outer-diameter gripper for the raw cast part and the
inner-diameter gripper for removing the bored part from the chucks.
The double hand used in a die casting process includes a biscuit
gripper and two expanding mandrel devices for holding the inserts.
A variety of insert mandrels accommodates different-size inserts.

In robot handling of bottles of photographic chemicals, the hand
is constructed of special acid-resistant stainless stell. Neoprene
pads prevent the bottles from slipping out of position. In automotive applications end-effector design is needed in loading and unloading of the piston machining line, assembled head palletization,
valve stuffing, and engine handling.

A special vacuum air double hand is used in grinding harrow
discs. The hand consists of a group of suction cups attached to
spring steel fingers set on an angle compatible with the possible
range of concavity of the discs. A separate pneumatic system for
each hand enables independent hand programming. In the aus-
shaping process of harrow disc manufacture, electric cartridge heat-
ers are used in each jaw to maintain a suitable temperature for the
spot quenching of disc contact areas. The discussion above is
summarized in Table 2.1.

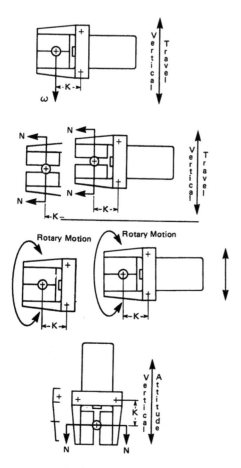

FIGURE 2.24   Principles of gripper configuration.   (Courtesy of
PHD, Inc.)

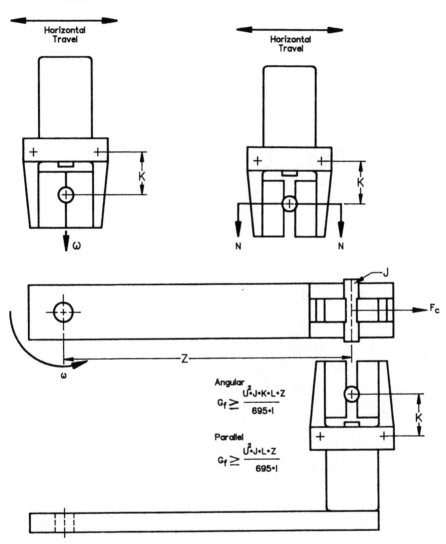

FIGURE 2.25 Further principles of robot configuration. (Courtesy of PHD, Inc.)

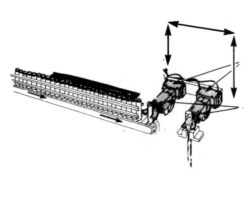

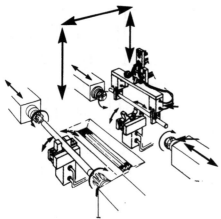

FIGURE 2.26   Grippers for various applications.   (Courtesy of
PHD, Inc.)

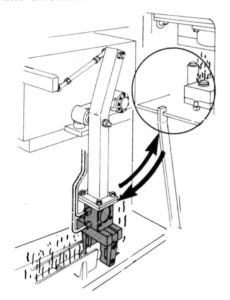

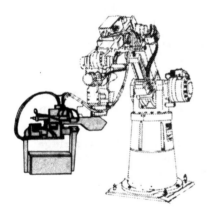

FIGURE 2.26 (Continued)

TABLE 2.1   Applications of Grippers

Application	Grippers
1. Palletizing rubber bales	Large vacuum cup with puncturing device
2. Manufacture of tapered plugs	Two grippers mounted at 180°
3. Production of forged diesel engine crankshafts	Two-forked hand with pneumatic toggle locking fingers
4. Transfer of differential case castings	Jaws with carbide-tipped inserts for gripping
5. Feeding billets through a two-cavity die forging press	Cam-operated parallel motion gripper
6. Chain link fabrication	Special grippers with carbide-tipped fingers
7. Processing gear blanks	Outer-diameter and inner-diameter double gripper
8. Robot handling bottles of photographic chemicals	Hand constructed of special acid-resistant stainless steel with neoprene pads
9. Grinding harrow discs	Set of suction cups attached to spring steel fingers set on an angle
10. Aus-shaping process of harrow disc manufacture	Each jaw with electric cartridge heater to maintain suitable temperature

REVIEW EXERCISES

1. Give the principal categories of grippers and describe two of them.
2. What are the possible kinematic arrangements that can be used to drive end effectors.
3. Describe universal grippers. How can the gripping force be improved in industrial applications?
4. What are dual grippers and why are they necessary?
5. Discuss the design factors that must be considered in the applications of end effectors.

6. Describe several applications of industrial robots and comment on the end effectors used for these applications.

FURTHER READING

Alvite, J., "Industrial End Effectors," *Proc. Robot 8 Int. Conf. Industrial Robots*, Detroit, June 1984, pp. 17.39–17.46.

Bracken, F. L., "Parts Classification and Gripper Design for Automatic Handling and Assembly," *Proc. 5th Int. Conf. Assembly Automat.*, Paris, May 1984, pp. 181–190.

Chelpanov, I. B. and Kolpashnikov, S. N., "Problems with the Mechanics of Industrial Robot Grippers," *Mech. Mach. Theory*, Vol. 18, No. 4, 1983, pp. 295–299.

Chen, F. Y., "Gripping Mechanisms for Industrial Robots," *Mech. Mach. Theory*, Vol. 17, No. 5, 1982, pp. 299–311.

Crawford, K. R., et al., "Intelligent Interchangeable Robotic End-of-Arm Tooling," *Proc. Robot 8 Int. Conf. Industrial Robots*, Dearborn, MI, June 1984, pp. 17.31–17.37.

Cutosky, M. R. and Kurokawa, E., "Grippers for an Unmanned Forging Cell," Carnegie-Mellon University Report No. CMU-RI-83-3, 1983.

Ennis, G., "Flat Workpiece Pickup," Department of the Air Force Report No. PAT-APPL-6195, 1983.

Engelberger, J. F., "Gripping Devices and Tools for Industrial Robots," *Werkstatt Betr.*, Vol. 114, No. 9, 1981, pp. 621–626 (in German).

Frohlich, W., "Grippers and Clamping Elements—Modular Elements for the Automation of Production Technology," *Industrial Robot*, Dec. 1979, pp. 195–198.

Jacobi, P. and Gnauck, G., "Design and Gripping Devices for Robots," *Maschinenbautechnik*, Vol. 30, No. 5, 1981, pp. 198–201 (in German).

Kato, I., *Mechanical Hands Illustrated*, P. B. Co. Ltd., Tokyo, 1982 (English editor Sadamanto, K.).

Kolpashnikov, S. N. and Chelpanov, T. B., "The Problem of Mechanics of Industrial Robot Grippers," *Proc. 4th CISM—IFToMM Symposium ROMANSY*, Warsaw, 1981.

Kolpashnikov, S. N., and Chelpanov, T. B., "Defining of the Scheme and the Parameters of an Industrial Robot Gripper by a Group of Specifications," *Proc. 11th ISIR*, Tokyo, 1981.

Konstantinov, M. S., "Jaw-Type Gripper Mechanism," *Proc. 5th ISIR*, Chicago, 1975, pp. 329–334.

Li, Y. and Zarrugh, M. Y., "Mechanical Design and Force Control of Manipulator Grippers," ASME Paper No. 83-DET-58, 1983.

Lundstrom, G., "Industrial Robot Grippers," *Industrial Robot*, Dec. 1973, pp. 72–81.

Manogg, H., "Concept for Versatile Handling of Parts of Industrial Robots," *Konstruktion (Berlin)*, Vol. 35, No. 6, 1983, pp. 239–245 (in German).

Mutter, R. F., "Effective Interfacing through End Effectors," *Proc. 13th Int. Symp. Industrial Robots* and *Robots 7*, SME, April 1983, pp. 4.1–4.11.

Nerozzi, A. and Vassura, G., "Study and Experimentaiton of a Multi-Finger Gripper," *Proc. 10th Int. Symp. Industrial Robots*, Milan, March 1980, pp. 215–223.

Neubauer, G., "Pneumatic Grippers," *Mach. Des.*, Nov. 25, 1982, pp. 69–71.

Podoloff, R. M. and Seering, W. P., "Design Considerations for a Robot End Effector for Automatic Assembly," *Proc. ASME Winter Annu. Meet.*, Nov. 1983, pp. 199–204.

Reed, C. K., "Examining the Use of Dual Grippers," *Robotics Today*, Vol. 5, No. 5, 1983, pp. 43–47.

Schafer, H. S. and Malstrom, E. M., "Evaluating the Effectiveness of Two-Finger Parallel Jaw Robotic Grippers," *Proc. 13th Int. Symp. Industrial Robots* and *Robots 7*, SME, April 1983, pp. 18.18–18.123.

Schmidt, I., "Flexible Moulding Jaws for Grippers," *Industrial Robot*, March 1979, pp. 24–26.

Siemens, K. J., "Constructive Solution Methods for Increasing the Feasibility of Tools for Handling Devices," *Fortschr. Ber. VDI Z. Reihe*, Vol. 2, 1983 (in German).

Suh, H. W., Cho, H. S., and Lee, C. W., "Grasping Force Control for a Robot Gripper Using Pneumatic On-Off Servomechanisms," *Proc. ASME Winter Annu. Meet.*, Nov. 1983, pp. 233–242.

Tella, R., Kelly, R., and Birk, J., "Contour Adapting Vacuum Gripper," *Proc. 10th Int. Symp. Industrial Robots*, Milan, March 1980, pp. 175–189.

Tur-Kaspa, Y. and Lens, E., "A Gripper for Ultra Thin Walled Tubes with a Built-in Force Sensor," *Sensor Rev.*, Vol. 3, No. 2, 1983, pp. 86–88.

Vranish, J. M., "Quick Change System for Robots," *Proc. Robot 8 Int. Conf. Industrial Robots*, Dearborn, MI, June 1984, pp. 17.74–17.83.

Wright, A. J., "Light Assembly Robots—An End Effector Exchange Mechanism," *Mech. Eng.*, July 1983, pp. 29–35.

3
End-of-Arm Tooling and Workpiece Positioners

In Chapter 2 we discussed various types of grippers used in robot applications. End-of-arm tools are devices such as welding guns, drills, and spray guns which are manipulated by a robot arm to perform specific tasks. In this chapter we will discuss a number of end-of-arm tooling components. First we will treat wrist mechanisms which help to manipulate the tools appropriately. Next we will provide a case study of quick-change wrist adaptors for improving the application of end-of-arm tools. Other robotic interchangeable tooling is discussed next. This is followed by a description of an aerospace drill. Finally, we will provide the features and application of workpiece positioners.

3.1 ROBOT WRISTS

Wrist mechanisms mounted at the end of a robot arm support the gripper or any other end-of-arm tooling. They may be driven by electric motors, hydraulic drives, or pneumatic drives. The mechanical design of wrists in industrial robots take several forms, including linkage drives, torque tubes, triordinate drive shafts, and hydraulic wrists. See Figures 3.1 through 3.4 for examples.

The wrist axes enable a robot to work in a variety of orientations and consist of the roll, pitch, and yaw axes. Roll is a rotation in a plane perpendicular to the end of the arm. Pitch is a rotation in a vertical plane through the arm. The third possible wrist motion, yaw, is defined as a rotation in a horizontal plane through the arm. The most common wrists in pick-and-place applications have two degrees of freedom for pitch and roll or yaw and pitch

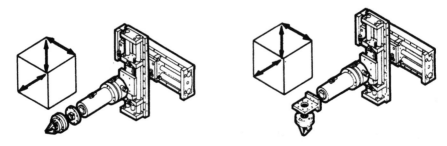

FIGURE 3.1 Fixed wrist configurations. (Courtesy of Mack Corporation.)

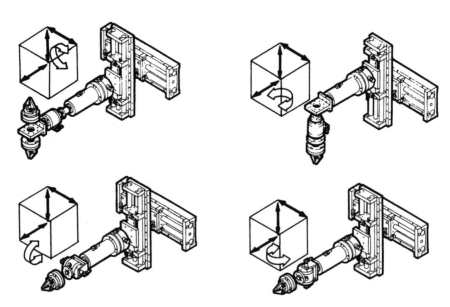

FIGURE 3.2 Wrist configurations with one degree of freedom. (Courtesy of Mack Corporation.)

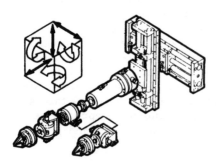

FIGURE 3.3 Wrist configuration with two degrees of freedom. (Courtesy of Mack Corporation.)

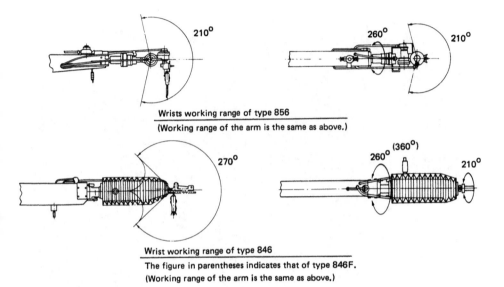

FIGURE 3.4 Working ranges of Tokico robot wrists. (Courtesy of Tokico America, Inc.)

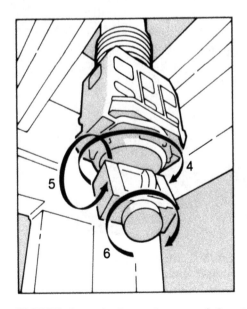

FIGURE 3.5 A three-degree-of-freedom wrist using Euler angles. (Courtesy of Cybotech Corporation.)

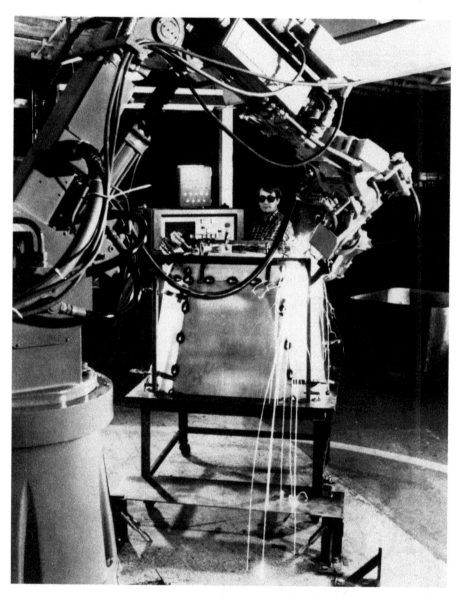

FIGURE 3.6 The Cincinnati Milacron T^3, with a three-degree-of-freedom wrist. (Courtesy of Cincinnati Milacron.)

orientations. For more complex applications, wrists with three degrees of freedom are needed to give roll, pitch, and yaw. Such wrists are used in many robot systems.

Wrists can be classified into two main types depending on the order of rotations to achieve the required orientation of the end effector. In the first type, the first rotation of the wrist is a roll about the axis of the third link of the robot arm, as shown in Figure 3.5. Typically this yields Euler angles for the orientation of the end effector. In the second type, the first rotation is a bend or yaw about an axis perpendicular to the axis of link three of the arm, as shown in Figure 3.6. Typical industrial robots have wrists belonging to one or the other of these two main types.

Standard off-the-shelf wrists allow modular additions of roll and yaw axis motions. Availability of pneumatic and programmable servo-operated models yields a wide range of selections for particular applications.

3.2 EOA QUICK CHANGE ADAPTORS*

EOA Systems, Inc. has developed a robotic Quick Change Adaptor tool designed to pick up and replace a variety of tools within a single work cell. The Quick Change Adaptor can be mounted on a variety of robot arms by utilizing standard interface plates which feature mounting holes designed to fit the bolt pattern of the specific robot. EOA Systems can modify a Quick Change Adaptor and design tool mounting plates around specific applications. See Figures 3.7 and 3.8.

The Quick Change Adaptor couples by fitting over the end-effector locking pin on the tool mounting plate. This allows users to mount their own tooling and make all electrical, fluid, and air connections. While fitting over the end-effector locking pin, the Quick Change Adaptor also locates on the indexing pin. As a result, the adaptor precisely couples with the tool mounting plate, thus aligning electrical and pneumatic ports. The indexing pin also prevents the male adaptor plate from rotating once coupled. The Quick Change Adaptor has a fail-safe locking device incorporated in the system which maintains a locked position when the air supply is removed.

The Model 225SP Quick Change Adaptor is designed exclusively for interchanging integral and nonintegral spot welding guns. This unique Quick Change Adaptor can quickly disconnect the 200-amp electrical contacts needed in spot welding applications while also

---

*Courtesy of EOA Systems, Inc.

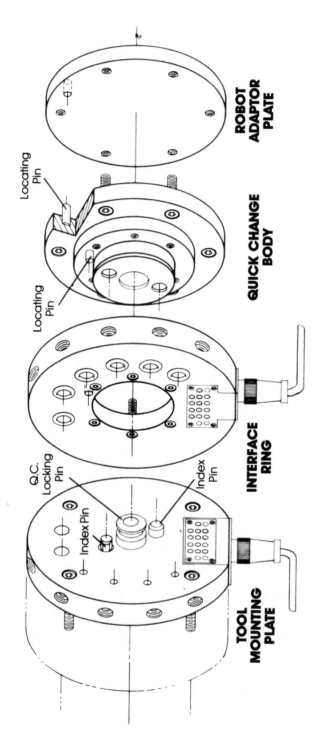

FIGURE 3.7 Components of Quick Change Adaptors. (Courtesy of EOA Systems, Inc.)

FIGURE 3.8 Quick Change Adaptors for a variety of tools within a single robotic work cell. (Courtesy of EOA Systems, Inc.)

FIGURE 3.8 (Continued)

quickly by disconnecting all pneumatic, fluid, and low-current connections to the integral spot welding guns. The capabilities of the
Model 225SP open many new horizons in robotic spot welding applications. Proven applications include:

1. Rapid change out of guns for maintenance operations.
2. Interchangeable guns on a re-spot line.
3. Redundant robots utilizing an interchangeable un system.

3.3 END-OF-ARM TOOLING

Typical tools that can be handled by general-purpose robots are

1. Stud welding head
2. Spot welding gun
3. Inert gas arc welding torch
4. Heating torch
5. Spray gun
6. Routing head, grinder, and sander
7. Ladle
8. Pneumatic nut runners

In a robot equipped with a stud welding head, the studs are fed
from a tubular feeder suspended from overhead. Stud welding heads
are relatively light, and their weight is rarely a significant limitation
since it is installed on robots of about 100-lbs capacity. An industrial robot can be equipped with a heating torch to bake out foundry molds. The heat is applied directly by playing the torch over
the surface. The bake-out time is shorter than the time it would
take to convey the molds through a gas-fired oven. A welding gun
attached to a robot can be used to place a series of spot welds on
flat and curved surfaces. Nut running, drilling, and countersinking
are operations well suited for robot applications. Mechanical grinders
are effective in increasing the locating accuracy and shortening the
positioning time.
    A router or grinder mounted on the wrist of a robot must typi-
cally be guided by a template for work in which a specific path must
be followed. In automated application of primers, paints, and ceramic
or glass frits, an industrial robot with an attached finishing gun is
effective for short or medium-length production runs. A single robot

FIGURE 3.9 End effector on the GCA over head gantry robot for
picking up automobile instrument panels. (Courtesy of GCA
Corporation.)

FIGURE 3.10 Spot welding car bodies as they move on conveyors. (Courtesy of Cincinnati Milacron.)

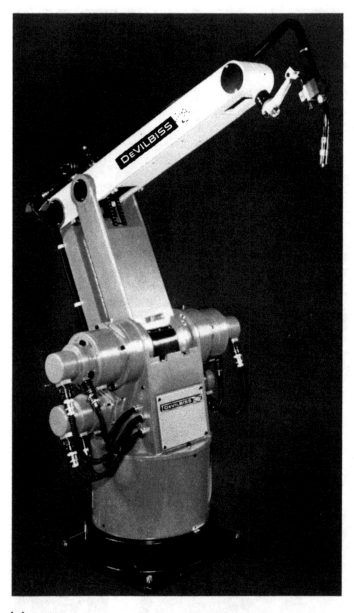

(a)

FIGURE 3.11 DeVilbiss robots with welding gun and spraying gun.
(a) The DeVilbiss EPR-1000 electrically powered arc welding robot;
(b) the DeVilbiss/Trallfa TR-4500 spray finishing robot. (Courtesy
of DeVilbiss Company.)

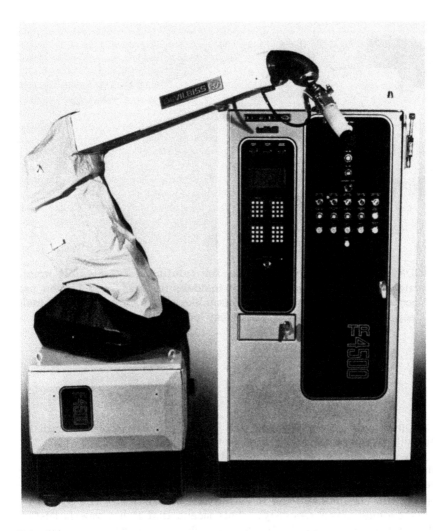

(b)

FIGURE 3.11   (Continued)

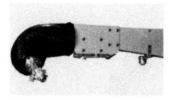

(a)

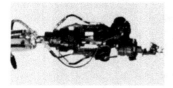

(b)

FIGURE 3.12 Robot end effectors for painting. (a) Flexible wrist
enables the robot to paint narrow insides, turnups, and hidden parts
of complicated workpieces. (b) Humanlike wrist allows painting inside
a deep boxlike structure. (Courtesy of Tokico America, Inc.)

equipped with a universal tool can handle several tools sequentially
in multiple operations on the same workpiece.

A robotic screwdriver is useful for flexible management of fastener installations with relatively little dependence on operator inter-
vention. A product which is well designed for robotic screwdriver
operations should have the following characteristics. The system
must provide error recovery. This means that if a screw is misfed,
the system will automatically clear the driver and ask for a new
screw. Also, if a screw is attempted and is rejected because of low
torque or insufficient depth, the robot can attempt to drive the same
hole again.

An example of such a robotic screwdriver uses sensors to provide reliable performance: (1) Fiber-optic sensors provide screw
presence and full depth insertion, (2) a pneumatic switch detects
torque condition, and (3) a Hall effect sensor and a magnetic target
detect part presence.

Figure 3.9 through 3.12 provide illustrations of the applications
of various end effectors.

3.4 DESCRIPTION OF EOA INTELLIGENT INTERCHANGEABLE ROBOTIC END-OF-ARM TOOLING*

Complex Drill

The complex drill embodies the sophistication necessary to perform the most difficult drilling applications. This high-technology drill can automatically drop off drill bits; can utilize up to 16 drill bits; can be assigned widely varied drilling parameters; has an 18-hp air motor with 400 lbs of thrust; and has the ability to countersink, drill to depth, peck drill, and use variable feed rates. The complex drill can detect broken drill bits, dull drill bits, and drill break-through. Also, for less sophisticated drilling applications, EOA Systems has developed the general-purpose drill. This manually adjustable drill cannot change out drill bits, but can detect drill bit breakthrough. The general-purpose drill is priced to be cost effective for less sophisticated drilling applications. See Figure 3.13 for examples.

Parts Measurement Gripper

The parts measurement gripper uses a high-resolution linear encoder to gauge machined or fabricated parts to less than 0.001 in., ensuring that strict quality controls are maintained in the work cell. In addition to its gauging function, the parts measurement gripper has a lifting capacity of 50 lbs and uses nonservo, air-actuated bolt-on fingers with a jaw travel of 4.0 in. Two pairs of infrared scanners are used for part detection in the fingers. The utility of the parts measurement gripper is significant for any machine load/unload applications where quality control is important. The parts measurement gripper allows the user to eliminate costly gauging systems and thus reduce overall cycle time in a machine load/unload application.

Transducerized Nutrunner

EOA Systems, Inc. offers a system of transducerized nutrunners covering a variety of torque requirements and fastener presentation methods. This transducerized nutrunner system provides the user with a large degree of flexibility when choosing the appropriate nutrunner for an application. Through the pre-processor, the user

---

*Courtesy of EOA Systems, Inc.

FIGURE 3.13   Various end-of-arm tooling systems.   (Courtesy of
EOA Systems, Inc.)

can format 4 sets of metric or english measuring units, specify
torque or turn of the nut mode, input 16 different tensions or
torque control "windows" per nutrunner.   The CRT display outputs
joint number, fastener size, thread per inch, actual torque and
tension, and target torque and tension.   In addition, the preprocessor can map the "joint signature" of a nut and, therefore,
can detect cross threading or any other fastener abnormality.   By
utilizing the tension control windows and joint signature the user
can ensure that all nuts are seated properly and precisely to the
user's specifications.   Finally, the nutrunner bracketry houses a
thrust mechanism to signal the robot to abandon the routine if the
nut is improperly seated and an inordinate amount of thrust is detected.   The EOA Systems nutrunner system is also available with
a hard copy printing function for quality control records.

Transducerized Screwdriver

In addition to offering the system of transducerized nutrunner, EOA
Systems, Inc. also offers a system of transducerized screwdrivers.
This system of transducerized screwdrivers covers a variety of
torque requirements.   The screwdriver can be programmed through
the pre-processor to perform up to 16 different torque control
parameters, thus ensuring screws are set precisely.   A variety of
screw presentation methods are available for use on this tool.

Deburr Tool

The software and sensing capabilities embodied in the EOA System
end-effector system give the deburr end effector a high degree of
real-time feedback. Position sensing through piezoelectric devices
enables this system to inform the robot to slow down or speed up
when a burr is detected. The user can format through the preprocessor the sensitivity of the burr detector and use discrete
input/output (I/O) to output a value proportional to burr size or a
specific signal, depending on the speed at which the robot arm is
to run. The adaptive feedback feature will allow the user to deburr
the part more completely.

3.5   AEROSPACE DRILL FEATURES*

Description of Software

The Aerospace Drill system represents a substantial upgrading in
robotic drilling technology. The ability to pick up and drop off
drill bits allows one drilling end effector to drill holes of many different sizes and dimensions. In addition, the sensory capabilities
of the Aerospace Drill and computer intelligence precisely control
the end effector to a very high degree.

The preprocessor communicates with the robot controller through
discrete I/O and coordinates motions of the end effector to execute
the commands of the controller. The user begins operation of the
Aerospace Drill by accessing the drilling parameters as defined by
EOA Systems. The video screen of the preprocessor will read as
follows:

1. Return to the previous menu
2. Change the perform now displayed
3. Feed rate of _____ inches/rotation
4. Drill to depth of _____ inches
5. Peck drilling stroke _____ inches
6. Average thrust _____ lbs
7. Drill bit detection threshold _____ lbs
8. Breakthrough detect threshold _____ lbs
9. Cage line to tip of bit _____ inches
10. Cage line to countersink _____ inches
11. Duplicate this perform to another
12. Perform a drilling operation using the parameters above
13. Alter drill characteristics

Please enter from 1 to 13.

---

*Adapted here by courtesy of EOA Systems, Inc.

Programming the drilling parameters is as easy as answering questions 1 through 13. It is important to remember that each hole drilled can have unique programmed drilling parameters. More important, each hole can be reprogrammed if the user wishes to change variables. Diagnostic checks are constantly being performed on the computer cards in the preprocessor and sensors on the drill to ensure that each is operating properly. The user can access the diagnostic routine to determine the viable operation of the important components of the system.

Mechanical Operation

The mechanical operation of the drill is initiated by extending the cage of the Aerospace Drill, which places the nosepiece inside the drilling template. The cage can be extended with up to 400 lbs of thrust (cage extend thrust is programmable to avoid damaging delicate parts) and continues to exert pressure on the part to keep the drill bit from walking on the part. Of course, cage extend thrust is an important consideration when using a "foot"-type nosepiece for freehand drilling.

After the cage is extended, the stepper motor begins to advance the drill motor and spindle down the ballscrew and toward the part. The stepper motor feeds the drill bit into the part at the feed rate programmed by the user through the preprocessor. The drill motor and bit are fed into the part until the drilling depth is reached or breakthrough occurs. After the drilling depth is achieved or breakthrough occurs. the drill motor and bit are rapidly retracted from the part. The preprocessor can determine drilling depth through the use of a linear encoder set inside the Aerospace Drill.

The drill bit and nosepiece are dropped off and picked up by using the stepper motor to drive the spindle down to the drop-off point. At the drop-off point the nosepiece and drill bit are dropped and the Aerospace Drill is ready to pick another drill bit and nosepiece. The Aerospace Drill is manufactured primarily from 2023 aluminum, with tooled steel used in critical areas. Use of 3/8-in. Thompson shafts gives the drill maximum support and minimizes deflection when the cage extends the drill nosepiece into the template. Thus the Aerospace Drill can remain rigid against the part when the cage is extended.

Operating Software

The operating software is stored in read-only memory (ROM), which is resident in the preprocessor. The software is composed of many subproblems written in "C" or assembler language, which run the various types of end effectors and allow the user to quickly alter

the work sequence under which any end effector operates. Each
end effector has an associated "ID code" from 1 to 16, depending
on the number of end effectors in the work cell. The preprocessor
checks for the proper ID before allowing the robot to use the end
effector, so the robot cannot use the wrong type of end effector for
a particular task. With each ID the user can program up to 16 (or
more if needed) different "performs" under which an individual end
effector can operate. For example, the Aerospace Drill might have
one perform which tells it to use a particular feed rate, depth control, and peck depth, while another perform specifies a completely
different format of operation such as retract after breakthrough,
use a lower feed rate, and do not peck drill. The robot controller
simply toggles one of 16 discrete I/O lines (which can be multiplexed
to save I/O) to initiate any one of the 16 different performs. The
number of drill bits to be used is not limited to 16. The customer
must specify the number of different formats to be used. The basic
concept behind these 16 performs is the same for all types of end
effectors, although the formats for each end effector differ depending on the type of end effector. The user can quickly modify the
formats at any time. Diagnostics are run continuously while in production. More thorough diagnostics can be run either under robot
control or by the user to precisely pinpoint a problem. In addition,
any end effector can be run manually while all relevant sensor information is being displayed in real time. The preprocessor can be
put into either manual, diagnostic, view-alter, or production mode.

Production Mode

Each end effector can be programmed to operate in a number of
ways. The robot control simply pulses one of the 16 "PERFORM"
lines going to the EOA preprocessor to indicate how it wishes the
end effector presently being held by the robot arm to operate (these
16 perform lines can be multiplexed). These perform lines are similar for all the various end effectors; how the EOA preprocessor
handles the toggling of one of these lines depends on which type of
end effector is currently being held on the robot arm and how the
user previously programmed the end effector to operate. Once the
user places the EOA Systems preprocessor into production mode, it
remains in that mode until either a fatal error occurs or the robot
or user hits the abort task line, at which time the system will go
back to the main menu. While in production mode the 16 perform
lines are constantly being monitored, as are the "PICK UP/DROP
OFF END EFFECTOR" or "PICK UP/DROP OFF DRILL BIT" control
lines. Actions taken when these lines are toggled are described
below.

Drop end effector. A check is made to see that quick change is not disabled. All solenoids and stepper motors are turned off, and the end effector is dropped by retracting the balls in the quick change. The "PROCEED" line to the robot controller is pulsed to tell the robot to move away from the dropped end effector; if the end effector is hung on quick change after 10 sec, the end effector is regrasped and the "TOOL CHECK" line is pulsed to tell the robot to move over for operator attention. If the uncoupled was successful, PROCEED is again pulsed to tell the robot to move to the next task (usually to pick up the next end effector).

Pick up end effector. A check is made to see that quick change is not disabled; if there is presently no end effector on the robot arm, then the quick change is uncoupled to prepare to pick up the end effector. The robot arm is moved to mating position after the preprocessor pulses the PROCEED line to indicate that all is well and that it is all right to get the end effector. When the preprocessor sees the identification code coming in from the quick change, the preprocessor grasps the end effector with the quick change. If a change in the identification code is not seen within several seconds the TOOL CHECK line is pulsed to send the robot into a tool check, since the robot probably became hung on the end effector while moving to grasp it. Once the robot has the end effector, the ID code from the newly acquired end effector is matched with the desired ID code from the robot controller. If the two do not match, the robot arm probably has acquired the wrong end effector. The robot is sent into a tool check since this is a fatal error which would be difficult for the robot to recover from without operator attention.

Drop drill bit. When this line is toggled, the drill will immediately drop the drill bit and nosepiece. After the drill has been dropped off, the PROCEED line to the robot controller will be toggled to tell the robot to proceed with the next task.

Get drill bit. The robot controller simply toggles one of the 16 perform lines to command the preprocessor to operate the Aerospace Drill using the proper parameters. The drill will drill to depth using electronic positive feed by reading spindle rpm; it can be set to feed at many feed rates (either inches per rotation or inches per minute) and will detect dull or broken bits and inform the robot controller and operator of several types of errors. The user may adjust the breakthrough detection and dull bit detection sensitivities. Error signals can be made to change a dull bit automatically (if the robot has an interrupt capability) or to send the robot into a tool check should the preprocessor detect an end-effector failure. In most cases the preprocessor will tell the robot controller to proceed with the next task by toggling the PROCEED

line. If an error occurs while drilling, the bit will be retracted
from the part automatically if doing so will not result in damage to
the part.

3.6   WORKPIECE POSITIONERS

Positioners expand the capabilities of a welding robot by maneuver-
ing the workpiece into the most efficient position for welding. For
parts which need welding on more than one side, positioning is mandatory. The working envelope of a robot is also extended by positioners which enable the robot to work on large weldments. Flat
welding provides a very effective position in which gravity aids
metal flow, improves weld quality, and speeds deposition. Positioners may also be used to transfer the weldments between the load
and unload station and the welding station. In this way productivity
of the robot is increased.

In conventional weldment positioners with two rotary axes, the
workpiece is clamped in a fixture to rotate continuously in a clock-
wise or counterclockwise direction. In the drop-center positioner,
the rotary table fits into a saddle below the level of the tilting
axis. This enables the workpiece to tilt about an axis close to its
center of gravity. A single-axis headstock-tailstock positioner is
ideally suited for long weldments. Some positioners have a powered
mechanism to elevate and adjust the table above the floor. See
Figures 3.14 through 3.18 for different workpiece positioners.

A dual positioner consists of two conventional positioners at opposite ends of a beam. An indexing axis under the beam center
rotates the positioner through 180° to present an unwelded workpiece to the robot and transfers a freshly welded workpiece to the
load-unload station. Dual positioners of the drop-center models are
used for relatively light workpieces in order to reduce the loading
on the long cantilever arms. With dual-table positioners, an operator stays behind a protective screen outside the working envelope
of the robot. Dual positioning takes up much floor space and the
extra axis may contribute to positioning errors.

There are two types of rotary-axis drives for positioners:
(1) indexing drives and (2) servo drives. Indexing drives assume
discrete positions which are determined by indexing stops. This
type of drive provides reasonably high speed and good position repeatability. Positioner drives that use limit switches tend to control
stop positions less accurately.

For complex assembly operations the available number of limit
switches may be inadequate to achieve all optimum seam attitudes.
The poor repeatability of limit switches may present a high probability or positioning error. These limitations of limit switches are

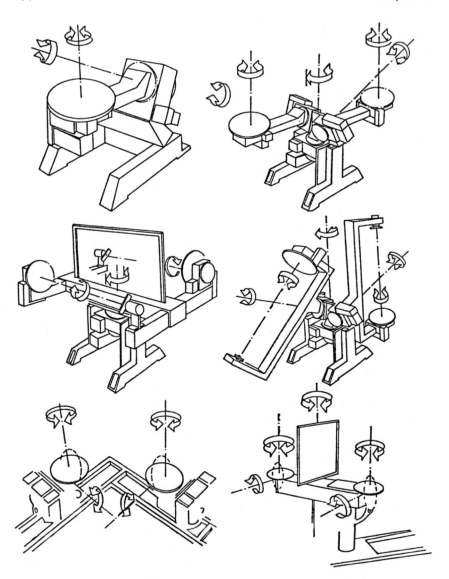

FIGURE 3.14 Workpiece positioners. (Courtesy of ESAB Robotic Welding.)

FIGURE 3.15   A welding robot with a workpiece positioner.
(Courtesy of ESAB Robotic Welding.)

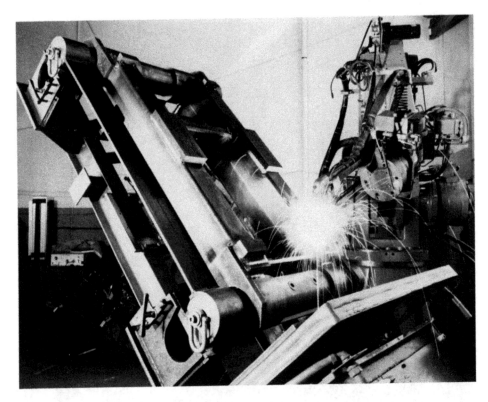

FIGURE 3.16   Cincinnati Milacron positioner-equipped T^3-566 robot
used to weld the upper arm assembly for its T^3-746 robot.   (Courtesy of Cincinnati Milacron.)

overcome by using a programmable weld positioner, which is easy to
teach and control.

Servo drive positioners integrate directly with the robot controller, permitting positioner axes to be programmed in the same way as
the robot axes. In some servo drive positioners, the positioner
movements can be coordinated with the torch motion. Servo drive
positioners tend to be slow and expensive. They are also rather
difficult to interface with robots not designed for arc welding.

Programmable workpiece positioners typically use resolvers interfaced with the robot control instead of cam-driven limit switches.
The resolvers measure the positions of the weld positioner axes. In
programming, the teaching pendant is used to drive the positioner
to the desired configuration. These locations are stored in robot
memory for automatic operation later on. Computer-controlled

FIGURE 3.17   Workpiece positioner and robot configuration. (Courtesy of Advanced Robotics Corporation.)

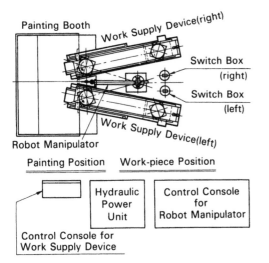

FIGURE 3.18   Work supply device in a robot painting system. (Courtesy of Tokico America, Inc.)

workpiece handling units increase welding speed, precision, and mobility. Because the workpiece turns around its own center of gravity, it can be placed in any attitude relative to the robot and still remain in the robot workspace. In some installations the workpiece handling unit may be mounted on the same foundation plate as the robot for stability and accuracy.

REVIEW EXERCISES

Using resource materials in your engineering library and manufacturer's handbooks, write brief and concise descriptions of selected end-of-arm tools. Indicate the application concepts and features of these end-of-arm tools. Your selection should include:

Robot wrist devices
Quick change adaptors
Robotic welding guns
Material removal tools
Nutrunners
Workpiece positioners

FURTHER READING

For further reading, refer to the reading list in Chapters 2 and 4.

4
End-of-Arm Sensors

4.1 TACTILE SENSORS

Many different types of tactile sensors based on conductive rubber, force-switchable diodes, or piezoresistive compounds have been tested. A force transducer using conductive silicone rubber is an effective touch- and force-sensing system. This device employs the changing electrical resistance of contact between an electrode and the conductive silicone rubber component as the two are forced together. The sensor can be made for a single sensitive point or in a matrix form to cover an area that is similar to an artificial skin. Force sensing is done at the interface between the robot and the object being manipulated. The device is responsive enough to drive an analog-to-digital converter over a reasonable range without amplification.

When welding beads on automotive bodies are to be ground by robots, variations in the beads and the location and curvature of the sheet metal resulting from production conditions complicate the control task. Tactile sensors attached to the robot arm give geometric feedback about the characteristics and the path of the welding bead. The control computer processes the measured data and calculates the ideal smooth surface and grinder paths. This enables the robot to remove welding beads on flat or convex surfaces and to smooth the surface automatically. The tactile sensors use strain gages which permit high measuring accuracy. A magnetoelastic and magnetostrictive force feedback sensor for robots and machine tools has been developed at the Naval Surface Weapons Center (NSWC). The basis of this tactile/force feedback sensor is the magnetostrictive effects of the materials. Force sensing with these materials may be accomplished by measuring the total force or the rate of change of the force on the sensor. The designs for robotic torque, grip, and slip modules are suitable for large industrial robots.

In edge-following operations such as welding and deburring, adaptive control enables a robot to adjust the program to compensate for deficiencies while working. Sensors and transducers produce feedback signals to adjust the robot speed and path of travel. This method of control is used to slow down the deburring operation or vary the weld path in arc welding.

Probing fingers with up to three degrees of freedom and proportional deflection transducers provide a basic tactile sensor in welding applications. Various control algorithms can be implemented to control the manipulation drives to follow the sensed contour characteristics.

The Tactile gripper shown in Figure 4.1 is a lightweight, force sensing gripper with a 10-lb lifting capacity and 3.25 in. of fingertip travel. The user can format the tactile gripper to grasp at various speeds until a certain position, grasping force, or part is detected. Also, the preprocessor can be commanded to report through discrete input/output (I/O) the following values: fingertip position, grasping force, or part presence. The tactile gripper is designed for use in lightweight gripping applications which require force feedback or fingertip positioning feedback.

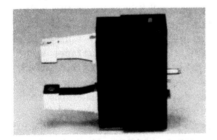

FIGURE 4.1   Tactile gripper.   (Courtesy of EOA Systems, Inc.)

4.2   PROXIMITY SWITCHES*

PHD magnetic reed limit switches can be used to provide an input
signal to a programmable controller or sequencer or to most control
mechanisms. The switches on the gripper can be adjusted to indicate a signal at any time during or at the end of jaw travel. The
signal can be used to determine whether the jaws have moved far
enough to grip a part or have moved too far, indicating that there
is no part in place. A magnetic piston must be specified on the
gripper to enable reed switches to be used.
      PHD Hall effect switches are specifically designed to provide an
input signal to microprocessors or computer-based controllers. They
are for use with DC only and provide a digital signal that can be
fed directly into a microprocessor. This switch is used with the
gripper in the same way as the reed switch, as it also indicates jaw position. A magnetic piston must be specified to enable Hall effect
switches to be used.
      PHD offers a solid-state sensor transducer along with a set point
module for sensing four or more positions throughout the jaw travel.
These signals can be used as inputs to a programmable controller
for indication of part presence, part sorting, or go/no go-type
gauging.
      The set point module allows independent adjustment of each
sensing position. See Figure 4.2 for sample proximity switches.

4.3   DESCRIPTION OF ASTEK GRIPPERS**

The Astek series GTP-45 pneumatic grippers are a family of compact,
quick-acting part pickup tools designed to hold parts rigidly during
automatic assembly operations. They are single-acting, have a
spring return, and are available with two or three fingers, either
actuated open or actuated closed. Tapped holes are provided in the
hardened steel fingers so that custom gripping jaws for specific
parts can be attached.
      The series GTP-45 grippers use a simple toggle linkage with a
large piston area to generate a gripping force at the end of the
fingers in excess of 2500 N (600 lbs) at 700 kPa (100 psi) pressure.
They are designed with a flange mount so that they may be readily
integrated with other tooling components, such as the Astek Accomo-
dator remote center compliance (Figure 4.3).

---

*Adapted by courtesy of PHD, Inc.
**Courtesy of Barry Wright Corporation.

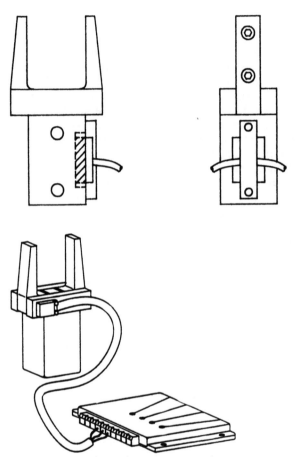

FIGURE 4.2   Proximity switches.   (Courtesy of PHD, Inc.)

     The optional finger-position sensor consists of a Hall effect
switch and an activating magnet. When the fingers grasp a part,
the piston is stopped in midstroke and the sensor remains inactive.
When the fingers do not grasp a part, the piston moves to the bottom of its stroke, activating the sensor. The sensor output signal
is a 12-V switch closure, which may be used by a machine controller
to indicate an error condition.
     Air pressure forces the piston down, moving the fingers on
their respective pivots. The spring pushes the piston back up after
pressure is removed. The sensor is tripped by the magnet when
the piston moves to its fully extended position.

(a)

(b)

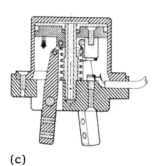

(c)

FIGURE 4.3   Features of Astek gripper. (Courtesy of Barry Wright Corporation.)

Sensoflex Gripper Pads   (Figures 4.4 and 4.5)

Durable nonslip elastomer gripper pads for industrial robots provide an efficient means of grasping workpieces. Elastomers are selected to operate over a broad temperature range and resist oils, other liquids, and corrosive elements. The elastomer is adhered to a metal plate for ease of attachment to a gripper. The pad is easily machined with a band saw to match custom applications, and holes may be drilled to provide quick attachment and removal. Two surface configurations are available, in several thicknesses. Gripper pads are also available without metal backplates for specialized applications.

FIGURE 4.4   Sensoflex gripper pads.   (Courtesy of Barry Wright Corporation.)

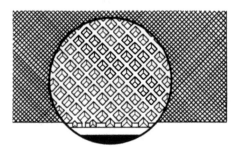

■ Knurled surface is designed for finer parts where cutting fluids are not involved.

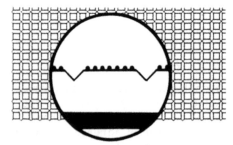

■ Waffle pattern allows drainage for cutting fluids while pebbled surface ensures a non-slip surface.

FIGURE 4.5   Operation of Sensoflex gripper pads.   (Courtesy of Barry Wright Corporation.)

Sensoflex Vacuum Gripper   (Figures 4.6 and 4.7)

The VG102-1 vacuum gripper is an innovative end-effector system designed for bin picking, assembly, and material transfer operations. The sensoflex vacuum gripper system consists of one gripper, P/N VG102-G1, and one control panel, P/N VG102-P1.

The vacuum gripper may be locked into a rigid or compliant mode and is also equipped with contact sense and grip sense. In a typical bin-picking application, the vacuum gripper enters the bin in the extended position. This allows the gripper to orient the vacuum cup to the outer contour of smooth objects randomly oriented at up to 60° from the gripper centerline. A signal is generated at initial contact and can be used to initiate a preprogrammed robot manipulator motion and/or turn on the vacuum source. A vacuum switch provides a signal when the part has been acquired which may be used to instruct the robot to leave the bin. After reaching a predetermined point, the robot may retract the flexible element and lock the object into a "home" position for accurate positioning and proper parts transport. The gripper control panel provides appropriate air supply/electrical interfacing.

Sensoflex Tactile System   (Figures 4.8 and 4.9)

The Sensoflex tactile system includes a low-profile, compliant tactile pad and an electronic interface device. The system provides data which can be used to determine gripping force, part position, and orientation, as well as a durable, compliant surface for part acquisition. A sense of touch is of particular importance for delicate

FIGURE 4.6   Sensoflex vacuum gripper system. (Courtesy of Barry Wright Corporation.)

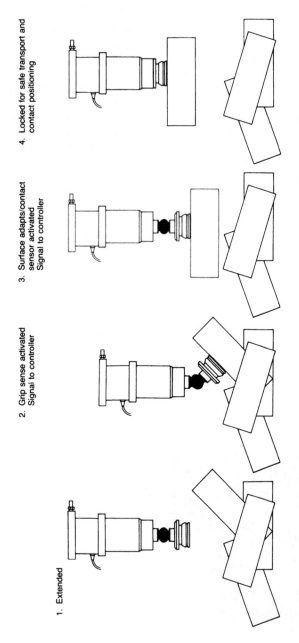

FIGURE 4.7   Operation of vacuum gripper.   (Courtesy of Barry Wright Corporation.)

FIGURE 4.8 Sensoflex tactile sensor. (Courtesy of Barry Wright Corporation.)

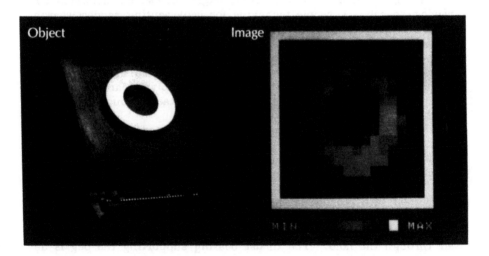

FIGURE 4.9 Typical graphic output of Sensoflex tactile sensor. (Courtesy of Barry Wright Corporation.)

assembly tasks in applications where discrimination between objects having different sizes, shapes, or weights is required.

The system operates by using conductive elastomer technology. The sensor yields an array of tactiles which are scanned on a row/column sequence. For example, row one is excited with a constant +5 V signal while reading the columns. Output from the interface device is an 8-bit digitized signal.

4.4   PASSIVE COMPLIANCE

Passive compliance denotes the ability of a robot to let external constraints modify the trajectory of the gripper. Passive compliant mechanisms have an adaptable output which permits self-correction in order to accommodate dimensional uncertainties in assembly tasks involving clearance fits, interference fits, and tooling interfaces. To cope with general linear and angular misalignments between a peg and a hole, passive compliant devices typically must have four to six degrees of freedom.

In peg-and-hole assembly tasks, alignment can be facilitated by random search. A simple technique for implementing random search is to vibrate a component with respect to the mating part. The vibration may be induced by special-purpose hardware fitted to the assembly head or to the gripper, or it can be performed by the robot itself. The peg can also be moved in a fixed but arbitrary pattern such as zigzag or spiral motion. In spiral search the robot essentially pokes for the hole and senses contact through the torques in the shoulder joint of the manipulator.

In heuristic searching, a rule of thumb or some commonsense strategy is used to govern the robot's search of the hole with the peg. A simple method is to advance the peg in the direction in which the peg tips when it makes contact with the hole. Another strategy involves creating an unbalanced force on the peg when it is misaligned with respect to the hole. The direction of this force is then used as the search direction. In some peg-hole alignment devices the required imbalance is created aerodynamically by blowing air through the hole. The thrust bearing supporting the peg is designed to enable the peg to respond appropriately to the imbalance. The axial thrust on the peg that opposes entry can be avoided by reversing the air flow. Programmable assembly experiments using air-bearing pads to provice some compliance have been demonstrated in industry. A specially designed hand performs chamfered pin-hole insertions by rocking the pin back and forth with the wobble angle plate. The limit of rocking is detected by means of limit switches, which are operated by the compression of springs in the wrist. The springs allow the pin to wobble with respect to the laterally working wrist. As presently implemented, this device will successfully assemble a close-fitting piston in a cylinder. Another version of this device uses position sensors to detect contact being made in the X and Y directions.

Remote center compliance (RCC) was developed and tested at the Draper Laboratories of the Massachusetts Institute of Technology (MIT). In response to a lateral positional error, the passive compliance system allows the part being assembled to realign laterally, provided the parts have chamfered edges and can be brought together with an initial positional error less than the dimensions of the chamfers. In response to an angular misalignment, the passive compliance allows the part to realign rotationally. A locking mechanism is used to hold the compliance device in a relatively rigid state while the parts to be assembled are being transported and positioned.

Possible uses of remote center compliance are in application areas involving clearance and interference fits and tooling interfaces. Examples of clearance fit applications are the assembly of bearings into housings, shafts into bearings, rivets into holes, and screws into threaded holes. Other examples include insertion of a forging into a die, insertion of an unmachined side of casting onto pallet locating pins, and precise assembly of delicate parts.

Applications with interference fits include the assembly of a nozzle into a housing, force fit of a shaft into a laminate stack, and snap fit of sheet metal parts. Tooling interface applications involve the placement of a tool into a machine tool changer, guide pins, and parts held in an escapement.

The remote center compliance device has been test run in research laboratories in the automotive industries for assembly operations such as throttle valve bushing assembly, intermediate servo piston assembly, and fuel pump assembly. The first evaluation is the insertion of an aluminum throttle valve bushing into a cast iron valve body. The cast chamfered ends of the aluminum bushing are inserted in the hole of the cast iron valve body, which is unchamfered. The intermediate servo piston is usually assembled automatically without an RCC device, but the close tolerances of mating parts and the galling problems make it difficult. The experimental evaluations demonstrate the usefulness of the RCC device, which does not produce jamming or wedging. The fuel subassembly contains a thermo disc valve which is press fitted to a sholdered tubular line. In manual assembly, several incomplete assemblies must be scrapped because of angular error during loading of the arbor press. The results obtained with an RCC device in the laboratory tests indicate that no jamming, wedging, or tilting is encountered with the assembly operations. These examples illustrate the applicability of a compliant device mounted on a robot arm for precision assembly in the automotive industry.

4.5   FEATURES OF REMOTE CENTER
COMPLIANCE DEVICES*

Automatic assembly of close-fitting parts can present costly problems.
If tolerance buildup is excessive, damaged parts, damaged assemblies, and damaged machines can contribute to waste and production
downtime.   Some of the increased productivity and reduced cost associated with automatic assembly machines is lost.

Remote center compliance devices provide the necessary give or
wrist action to allow close-fitting parts to be assembled efficiently.
They prevent jamming, wedging, galling, or similar damage.   In
machining operations which require the toll to conform to a template
or premachined location, the RCC accomodates tolerance buildup.

Typical applications are in operations such as assembly of bearings into housings, shafts into bearings, gears onto spline shafts,
rivets into holes not precisely located, screws into threaded holes,
delicate part assembly, and press fit.

Other operations in which RCCs are used include automatic gauging, mold alignment, positioning on assembly lines, general machining, template drilling, template routing, and reaming.

Remote center compliance devices have a number of advantages
including:

1.   Facilitate automatic assembly in more applications
2.   Allow positioning tolerances to be relaxed
3.   Reduce downtime of assembly equipment by reducing wedging,
     jamming, and damaging of parts
4.   Reduce the scrap rate normally caused by high forces or moments
     on parts
5.   Reduce shaving and eliminate galling of press fits
6.   Replace floating heads, which have breakaway force and wear
     disadvantages
7.   Eliminate expensive electronics normally required for precise
     positioning

The operation of an RCC is as follows.   A compliant center or
elastic center is the point about which rotation will occur when a
moment is applied.   When a lateral force is applied at the elastic center, only lateral translation occurs.   Projecting this point to the
leading edge of a part to be inserted is the basis for the RCC device performing its function.   To facilitate this, a chamfer is required.   The part is held at the end away from the mating part.
The RCC device projects the compliant center to the leading edge of

---

*Adapted by courtesy of Lord Corporation.

the part. When the part is inserted, lateral forces caused by interference produce translational reaction or motion through the RCC device, relieving the interference. If the interference causes a moment
on the part, a rotational reaction or motion will also relieve the
interference.

This combination of reactions minimizes insertion forces and jamming during assembly or insertion. Two characteristics make this
device effective: controlled elastic center projection (remote center
compliance) and controlled flexibility (stiffness).

Control is accomplished with laminated elastomer and metal shim
elements. In compression these elements are much stiffer than in
shear. This high ratio of compression to shear spring rate facilitates
elastic center projection and controlled flexibility. By changing the
elastomer, number of shims, and part geometry, the performance of
the RCC device can be altered to meet application needs.

Let us now describe how the assembly or insertion takes place
when an RCC is used (Figures 4.10 and 4.11). Lateral error in
position between the hole and shaft exerts a horizontal force on the
leading end of the shaft as a result of the chamfer. Acting approximately through the elastic center, the horizontal force causes the
shaft to translate laterally into the hole, permitting easy insertion.

Now let us suppose that the axis of the hole is not parallel to
the axis of the shaft. Positioning itself laterally, the shaft will

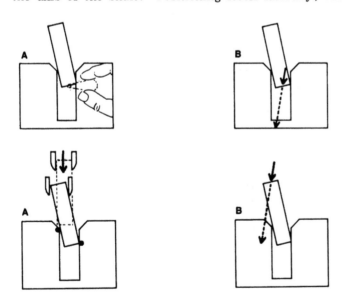

FIGURE 4.10   Principles of RCC action.   (Courtesy of Lord
Corporation.)

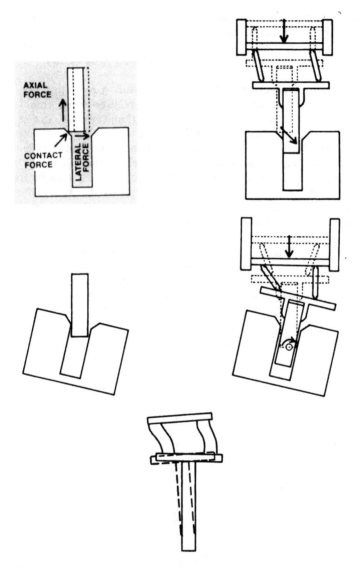

FIGURE 4.11   Operation of RCC devices.   (Courtesy of Lord Corporation.)

enter the hole. However, the leading edge will contact one side of the hole and the edge of the hole will contact the other side of the shaft, thus causing a moment. Rotation about the compliance center will allow the shaft to line up with the hole and be inserted easily. By combining the two modes of freedom, a usefully compliant device produces the desired performance results with one set of three flexing elements. Each has shear-to-compression spring rate characteristics which combine to project the center of compliance as required.

4.6 DESCRIPTION OF ASTEK ACCOMMODATOR

The Astek Accommodator remote center compliance device is a unique system which compensates for misalignment between part during automatic assembly operations. It allows an assembly machine or robot to compensate for positioning errors due to machine inaccuracy, parts vibration, and fixturing tolerances and thus minimizes the assembly forces and the possibility of parts jamming.

The AST-75, ASP-85, and AST-100 compliance devices are designed around a set of six elastomeric shear pads, which make the devices stiff in compression yet relatively soft in shear. This allows the Accommodator unit to move from side to side during an assembly operation to correct for both lateral and out-of-square misalignment. The AST-75 is a unit with a small diameter for applications where space is at a premium. The ASP-85 is light in weight and is provided with the PUMA mounting flange for its machine-side interface. The basic AST-100, the largest standard unit, can be fitted with optional antirotation pins so that torque can be transmitted.

These devices have the following features:

1. They compensate for lateral, out-of-square, and torsional misalignment.
2. They are stiff in compression for insertion operations.
3. Elastomeric shear pads provide inherent damping and self-centering.
4. Mechanical stops protect against overload in all directions.
5. A choice of center-of-compliance projection is available for different-size parts.
6. Alternative shear pad elastomers provide high- or low-stiffness units and harsh environment operation.

---

*Courtesy of Barry Wright Corporation.

FIGURE 4.12   Astek Accommodator RCC. (Courtesy of Barry Wright Corporation.)

Forces generated by mating parts cause elastomeric pads to deflect in shear. This allows parts to translate and/or rotate, compensating for misalignment. Rotation is about a point remote from the unit, called the center of compliance. See Figures 4.12 and 4.13 for an example and an operation diagram.

Center-of-Compliance Projection

The center of compliance is the point in space about which out-of-square rotation occurs. To minimize the forces generated during assembly, the projection distance should be chosen to place the center of compliance at the point of initial contact of the parts.

Elastomer Selection

Natural rubber provides the best mechanical properties but is susceptible to environmental degradation. Silicone rubber provides compatible performance at some sacrifice in hysteresis. Neoprene is environmentally robust and is the elastomer of choice for all but the softest units.

Stiffness

The Accommodator design allows a trade-off between the lateral and rotational stiffness values for a given projection distance and elastomer type. The units may be chosen to balance these two parameters.

Units may also be optimized for minimum lateral or minimum rotation stiffness.

An experimental setup at the MIT Artificial Intelligence (AI) Laboratory has been used to study robotic precise assembly tasks using a wrist sensor. The level of performance of the experimental system allows such basic actions as putting a peg into a hole, screwing a nut on a bolt, and picking a thin washer from a flat table. A number of parts including two ball bearings, several spacers, a threaded shaft, and a nut were assembled by a combination of strategies. The motion constraints in wrist force control described in a suitable coordinate system in terms of the principal axes are matched to force sensing elements mounted on the wrist. The differential motions in the constraint system are matched to individual joint motions. For each compliance, a separate force sensing element controls each joint. Changes in the coordinate system matrix compensate for the nonorthogonal joint motion.

Both the position and the contact forces generated at the hand can be controlled by a hybrid system. A wrist-mounted force sensor provides data for the implementation of this scheme. The controller incorporates two sets of closed loops for force and position control, which act together to achieve compliant control. Because the transfer function from the actuator sensor is different for force and position control, separate control laws are required. To obtain optimal results, feedforward compensators for arm dynamics and gravity loading must be incorporated in addition to error compensation mechanisms.

In an active adaptable compliance wrist (AACW) assembly, passive and active accommodation based on force feedback are combined. A

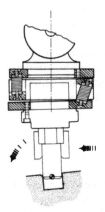

FIGURE 4.13   Operation of Astek Accommodator RCC. (Courtesy of Barry Wright Corporation.)

special end effector consists of a universal wrist with five degrees of freedom provided with independently software-controllable compliances and saturation levels. In this device the force sensor is incorporated in the wrist itself. In the assembly of closely fitting parts, the adaptable wrist provides compliance during the initial approach and the final insertion operation.

Another wrist sensor method is based on a so-called free-joint method. The "free joint" is defined as the joint which is most sensitive to motion in a required direction. It is the first joint to move if the force is slowly increased from zero in the free direction. A hybrid controller uses the free-joint concept to provide compliance by assigning individual joints for each degree of compliance and driving the remaining joints in a position servo mode. This method is a differential approximation to the compliance because only a subset of manipulator joints is force-controlled. Correction for resulting position errors is made on subsequent servo cycles. Draper Laboratories developed the Instrumented Remote Center Compliance (IRCC) as a wrist-mounted sensor that combines the attributes of passive compliance and active sensory feedback devices. The IRCC consists of a basic RCC and instrumentation capable of measuring the deflections of the compliant device in one or more degrees of freedom. The resulting compliant structure exhibits the advantages of both passive and active accommodation. A three-degree-of-freedom IRCC has been tested on the wrist of a standard Unimation PUMA industrial robot. Sensory information from the IRCC is in the form of a vector quantity which can be interpreted for touch or contact detection, position/angle measurement, and force/torque measurement. The IRCC has been tested in tasks requiring active accommodation such as chamferless insertions, edge following, seam tracking, and self-learning.

In locating the components in machining and gauging stations of a small-batch flexible machining cell, a compliant sphere assembly with automatically adjustable springs may be used. This is a general-purpose wrist unit for industrial manipulators which carry out assembly tasks that encompass a wide range of payloads. The device is instrumented with Linear Variable Differential Transducers (LVDT) that make it possible for the robot controller to adjust the arm.

4.7 DESCRIPTION OF ASTEK SIX-AXIS FORCE SENSOR WITH ONBOARD MICROPROCESSOR*

The Model FS6-120A six-axis force sensor is an integrated measuring instrument which includes the transducer structure, strain gages, and instrumentation electronics necessary to completely measure the

---

*Courtesy of Barry Wright Corporation.

six components of force that may be applied to the sensor body by an arbitrary load. The unit contains a built-in microprocessor which can resolve in real time the six force components about any specified coordinate system (Figures 4.14 and 4.15).

Output from the unit is an RS-232C-compatible serial bit stream which contains measurements of the instantaneous forces acting on the sensor body. A comprehensive input command language is interpreted by the onboard processor, allowing the operator to modify the coordinate system and other operating parameters from an external terminal or computer. An analog output port is provided for direct connection to a chart recorder or other analog readout device.

The electronic hardware is contained on two printed-circuit boards located within the sensor housing. Interconnection to the sensor is by a shielded twisted-pair cable, with separate lines for DC input power, RS-232 I/O, and alalog output and a discrete digital line for overlimit status.

Interface Considerations

The FS7-120A software system is designed to interact with either a stand-alone CRT terminal or a computer I/O port. In either case, the onboard processor interprets a simple American Standard Code for Information Interchange (ASCII) command set, allowing the user to set the various sensor operation parameters and control the output data stream. Data output is available in a number of formats, including an ASCII output format and a more compressed 16-byte

FIGURE 4.14   Astek six-axis force sensor.   (Courtesy of Barry Wright Corporation.)

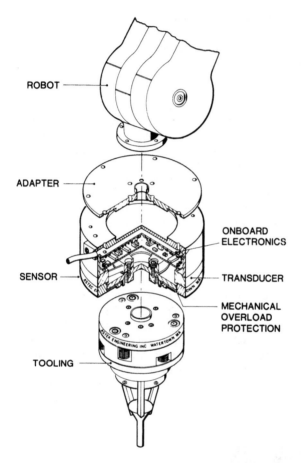

ROBOT

ADAPTER

ONBOARD
ELECTRONICS

SENSOR

TRANSDUCER

MECHANICAL
OVERLOAD
PROTECTION

TOOLING

FIGURE 4.15   Components of force sensor.   (Courtesy of Barry
Wright Corporation.)

binary format.   Also available are separate posts for analog data
and overlimit signaling.

Data Sampling Rate

The sensor output data rate is determined by both the operating
baud rate and the selected data output format and may range from
1 to 480 Hz.   To preserve the integrity of the sampled data signals,
the output bandwidth is appropriately limited by a programmable
digital filter; this guarantees that servo systems based on the sensor
output will not be affected by low-frequency aliasing components.

Sensor Bias

A bias, or DC offset, vector is subtracted from each set of analog-to-digital (A/D) data readings. This allows the sensor to compensate for gravity loads and any tooling which may be added to the sensor. This bias vector is reset by command input, which typically is done prior to any data gathering operation.

Input commands take the form of two-character ASCII mnemonics followed by optional numeric parameters; they are broken down into groups as follows:

Output select
Reset bias
Frame select
Limits select
Passband select
Threshold select
Baud rate select
Echo control
Configuration display
Self test
Miscellaneous functions

To summarize, the ASTEK system illustrates how a microprocessor can be used to enhance the operation of a force sensor.

REVIEW QUESTIONS

Prepare a class project report on selected robot sensors as assigned by the instructor. Construct a comparative chart to show the differences and similarities between these sensors. Discuss the impact of sensors on robot applications.

FURTHER READING

Bollinger, J. C., "The Use of Tactile Sensing for the Guidance of a Robotic Device for Welding," *Proc. 1st Int. Conf. Robot Vision Sensory Systems*, Stratford-upon-Avon, UK, April 1981, pp. 193–204.

Cutkosky, M. and Wright, P. K., "External Position Control of Industrial Manipulators," *Proc. 2nd Int. Conf. Comput. Eng.*, San Diego, Vol. 2, Aug. 1982, pp. 113–118.

Cutkosky, M. R. and Wright, P. K., "Position Sensing Wrists for Industrial Manipulators," *12th Int. Symp. Industrial Robots*, Paris, June 1982.

Fazio, T. L., Whitney, D. E., et al., "Feedback in Robotics for Assembly and Manufacturing," *Proc. 9th Conf. Production Res. Technol.*, NSF/MEA-81013, Ann Arbor, MI, Nov. 1981, pp. G1-G8.

Helms, D., "Enkwicklung eines sechs-Komponenten Kraft-Moment-Aufnemers," *Feinwerktechnik Messtechnik*, 85, 1977.

Inoue, H., "Force Feedback in Precise Assembly Tasks," MIT Memo AIM 308, August 1974.

Krouse, J. K., "Compliant Mechanisms - A New Class of Mechanical Devices," *Mach. Des.*, Jan. 24, 1980, pp. 86-90.

Nitzan, D., "Assessment of Robotic Sensors," *Proc. 1st Int. Conf. Robotic Vision Sensory Controls*, Stratford-upon-Avon, UK, April 1981.

Paul, R. and Shimano, B., "Compliance and Control," *Proc. JACC*, San Francisco, 1976.

Paul, R. L. and Wu, C. H., "Manipulator Compliance Based on Joint Torque Control," IEEE Conference on Decision and Control, Albuquerque, NM, Dec. 1980.

Presern, S., Spegel, M., and Ozimek, I., "Tactile Sensing System with Sensory Feedback Control for Industrial Arc Welding Robots," *Proc. 1st Int. Conf. Robot Vision Sensory Systems*, Stratford-upon-Avon, UK, April 1981, pp. 205-214.

Pugh, A., Heginbotham, W. B., and Page, C. J., "New Techniques for Tactile Imaging," *Radio Electron. Eng.*, Vol. 46, No. 11, Nov. 1976, pp. 519-522.

Pugh, A., Heiginbotham, W. B., and Sato, N., "A Method for Three Dimensional Part Identification by Tactile Transducer," *Proc. 7th Int. Symp. Industrial Robots*, Tokyo, 1977, pp. 587-593.

Purbick, J. A., "A Force Transducer Employing Conductance Silicone Rubber," *Proc. 1st Int. Conf. Robot Vision Sensory Controls*, Stratford-upon-Avon, UK, April 1981, pp. 73-80.

Simons, J., "Force Feedback in Robot Assembly Using an Active Wrist with Adaptable Compliance," Ph.D. Dissertation, Katholieke Universiteit, Leuven, 1980.

Stute, G. and Erne, H., "The Control Design of an Industrial Robot with Advanced Tactile Sensitivity," *Proc. 9th Int. Symp. Industrial Robots*, Washington, DC, 1979, pp. 519-527.

Van Brussel, H. and Simons, J., "The Adaptable or Compliance Concept and Its Use for Automatic Assembly by Active Force Feedback Accommodations," *9th Int. Symp. Industrial Robots*, Washington, DC, 1979, pp. 167-181.

Van brussel, H. and Simons, J., "Adaptive Assembly," *Proc. 4th Brit. Robot Assoc. Conf.*, IFS, 1981.

Van Brussel, H. and Simons, J., "Automatic Assembly by Active Force Feedback Accommodation," *Proc. 8th Int. Symp. Industrial Robots*, 1978.

Van Brussel, H. and Simons, J., "A Self Learning Robot for Automatic Assembly," *Proc. 1st Int. Conf. Assembly Automation,* Brighton, 1980.

Vranish, J. M., "Magnetoelastic Force Feedback Sensors for Robots and Machine Tools," *Proc. Robot VI Conf.,* Detroit, March 1982, pp. 492-522.

Wang, S. and Will, P., "Sensors for Computer Controlled Mechanical Assembly," *Industrial Robot,* March 1978, pp. 9-18.

Watson, F. P. and Drake, S. H., "Pedestal and Wrist Force Sensors for Assembly," *Proc. 4th Int. Symp. Industrial Robots,* 1975

Compliance

Craig, J. J. and Raibert, M. H., "A Systematic Method for Hybrid Position/Force Control of a Manipulator," *IEEE COMSAC Conf.,* Chicago, Nov. 1979.

Drake, S. H., "The Use of Compliance in a Robot Assembly System," presented at the IFAC Symposium on Information-Control Problems in Manufacturing Technology, Tokyo, Oct. 1977.

Drake, S. H., Watson, P. C., and Simunovic, S. H., "High Speed Robot Assembly Precision Parts Used in Compliance Instead of Sensory Feedback," *Proc. 7th Int. Symp. Industrial Robots,* Tokyo, 1977, pp. 87-97.

Goryachko, V. I., et al., "Vacuum Method of Automatic Assembly," *Russ. Eng. J.,* Vol. 57, No. 1, 1977, pp. 61-62.

Goto, T., Onoyama, T., and Takeyasu, K., "Precise Insert Operation by Tactile Controlled Robot HI-TI-HAND Expert 2," *Proc. 4th Int. Symp. Industrial Robots,* Tokyo, Nov. 1974.

Groome, R. C., Jr., "Force Feedback Steering of a Teleoperator System," M.Sc. Dissertation, Cambridge, MA, Aug. 1972.

Hewit, J. R. and Burdess, J. S., "Fast Dynamic Decoupled Control for Robotics, Using Active Force Control," *Mech. Mach. Theory,* Vol. 16, No. 5, 1981, pp. 535-542.

Karelin, N. M. and Girel, A. M., "The Accurate Alignment of Parts for Automatic Assembly," *Russ. Eng. J.,* Vol. 47, No. 9, 1967, pp. 73-76.

Karelin, N. M. and Girel, A. M., "Automation of Component Assembly Operations in a Rotating Magnetic Field," *Russ. Eng. J.,* Vol. 54, No. 10, 1974, pp. 53-56.

Mason, M. T., "Compliance and Force Control for Computer Controlled Manipulators," MIT Artificial Intelligence Laboratory Technical Report 515, April 1979.

McCallion, H. and Wong, P. C., "Some Thoughts on the Automatic Assembly of a Peg and a Hole," *Proc. 4th World Congress on the Theory of Machines and Mechanisms,* Inst. Mech. Eng., London, Sept. 1975, pp. 347-372.

McCallion, H., Alexander, K. V., and Pham, D. T., "Aids for Automatic Assembly," *1st Int. Conf. Assembly Automation,* Brighton, UK, March 1980.

Nevins, J. L. and Whitney, D. E., "The Force Vector Assembler Concept," *1st Int. Conf. Robots Manipulators,* Udine, Italy, Sept. 1973.

Nevins, J. L., et al., "Exploratory Research in Industrial Assembly Part Mating," Charles Stark Draper Laboratory, Cambridge, MA, Seventh Progress Report for the NSF, Feb. 1980.

Raibert, M. H. and Craig, J. J., "Hybrid Position/Force Control of Manipulators," *ASME J. Dynamic Syst. Measurement Control,* Vol. 102, June 1981.

Salisbury, K., "Active Stiffness Control of a Manipulator in Cartesian Coordinates," *Proc. 19th IEEE Conf. Decision Control,* Albuquerque, NM, Dec. 1980.

Savishchenko, V. M. and Beaspalov, V. G., "The Orientation of Components for Automatic Assembly," *Russ. Eng. J.,* Vol. 45, No. 5, 1965, pp. 50–52.

Schneider, T., "Entwicklung aund Bau einer sich Selbst Steurnden Roboterhand", Institut für Industrielle Fertigung and Fabrikbeirieb, Universitat Stuttgart, March 1976.

Shimano, B., "The Kinematic Design and Force Control of Computer Controlled Manipulators," Stanford A.I. Lab. Memo. 313, March 1978.

Watson, P. C., "The RCC Concept and Its Application to Assembly Systems," *IFAC Symp. Information-Control Problems in Manufacturing Technol.,* Tokyo, Oct. 1977.

Watson, P. C., "The Remote Center Compliance System and Its Application to High Speed Robot Assemblies," Society of Manufacturing Engineers, Paper No. AD77-718, 1977.

Whitney, D. E., "Force Feedback Control of Manipulator Fine Motion," *Proc. JACC,* San Francisco, 1976.

Whitney, D. E., "Resolved Motion Rate Control of Manipulators and Human Prostheses," *IEEE Trans. Man-Machine Syst.,* No. 2, June 1969, pp. 44–53.

Wohlert-Jensen, C. H. H., "Techniques for Automatic Assembly," Ph.D. Thesis, University of Cantebury, New Zealand, 1978.

Yakhimovich, A. A., et al., "Automatic Assembly of Components by the Jet Method," *Russ. Eng. J.,* Vol. 50, No. 6, 1970, pp. 58–63.

Sensors

Amat, J., et al., "Sensing Systems in Industrial Robots," *Mundo Electron.,* No. 108, 1981, pp. 59–65 (in Spanish).

Cassimis, R., "Robot Sensor: Inherent Problems and Future Developments," *Autom. Instrum.,* Vol. 29, No. 4, 1981, pp. 299–301 (in Italian).

Dillman, R. and Faller, B., "Force Moment Sensor System for Industrial Robots," *Electronik*, Vol. 31, No. 18, 1982, pp. 89–85 (in
German).

Foulc, J. N. and Lopez, F., "Initiation to Robotics. IV. Sensors,"
Le Nouvel Automatisme, Vol. 26, No. 19, 1981, pp. 53–60 (in
French).

Hudson, C., "Sensors and Robots, Mechanical Actuation to No-Touch
Sensors," *Proc. Conf. Record Industry Applications Soc. IEEE-
IAS Annu. Meet.*, Philadelphia, Oct. 1981.

Weaver, J. A., "Sensing Systems for Robots," *Proc. 4th Brit. Robot
Assoc. Annu. Conf.*, Brighton, UK, May 1981, pp. 191–201.

Tactile Sensors

Benjamin, H. L., "The Development of a Production Robot Tactile
Position Sensor," *Proc. 13th Int. Symp. Industrial Robots* and
Robots 7, Chicago, April 1983, Vol. 2, pp. 18.57–18.76.

Chalupa, V., Marik, V., and Volf, J., "Tactile Matrix for Shape
Recognition," *Technol. Methodol. Adv. Measurement, Proc. 9th
IMEKO Cong. Int. Measurement Conf.*, May 1982, West Germany,
Vol. 1, pp. 339–348.

Checeinski, S. S. and Agrawal, A. K., "Magnetoelastic Tactile Sensor," *Proc. 3rd Int. Conf. Robot Vision Sensory Controls*,
Cambridge, MA, Nov. 1983, pp. 333–340.

Christ, J. P. and Sanderson, A. C., "Prototype Tactile Sensor
Array," Carnegie-Mellon University Interim Report No. CMU-RI-
TR-82-14, 1982.

Davies, J. B. C., "Carbon Fibre Sensors," *Proc. 4th Int. Conf.
Robot Vision Sensory Controls*, London, Oct. 1984, pp. 295–302.

Deer, D. J., "3D Tactile Sensing," *Sensor Rev.*, Vol. 3, No. 2,
1983, p. 59.

Gaston, C. and Lozano-Perez, T., "Tactile Recognition and Localisa-
tion Using Object Models: The Case of Polyhedra on a Plane,"
MIT Report No. AIM-705, 1983.

Gindy, S. S., "A New Concept for Strain Gauge Tactile Sensing,"
Proc. 8th Int. Conf. Industrial Robots, Detroit, June 1984,
pp. 21.19–21.32.

Grimson, E. E. and Lozano-Perez, T., "Model-Based Recognition and
Localization from Sparse Range or Tactile Data," Memo. Report
No. AI-M-738, 1983.

Hackwood, S., Beni, G., and Nelson, T. J., "Torque-Sensitive Tactile Array for Robotics," *Proc. 3rd Int. Conf. Robot Vision
Sensory Controls*, Cambridge, MA, Nov. 1983, pp. 363–370.

Harmon, L. D., "Automated Touch Sensing: A Brief Perspective
and Several New Approaches," *Proc. Int. Conf. Robotics*,
Atlanta, GA, March 1984, pp. 326–331.

Hollerbach, J. M., "Workshop on the Design and Control of Dexter-
ous Hands," Report No. AI-M-661, Nov. 1981.

Julliere, M., et al., "Tactile Arm for Mobile Robot," *Le Novel Automatisme*, Vol. 21, April 1981, pp. 62–66.

Konishita, G., et al., "Development and Realisation of a Multi-Purpose Tactile Sensing Robot," *Dev. Robotics*, 1983, pp. 185–196.

Konoshita, G., Hajika, T., and Hattor, K., "Multifunctional Tactile Sensors with Multielements for Fingers," *Proc. Int. Conf. Advanced Robotics*, Sept. 1983, Part 1, pp. 195–202.

Mott, D. H., Lee, M. H., and Nicholls, H. R., "An Experimental Very High Resolution Tactile Sensory Array," *Proc. 4th Int. Conf. Robot Vision Sensory Controls*, London, Oct. 1984, pp. 241–250.

Raibert, M. H. and Tanner, J. E., "A VLSI Tactile Array Sensor," *Proc. 12th Int. Symp. Industrial Robots/6th Int. Conf. Industrial Robot Technol.*, Paris, June 1982, pp. 417–425.

Ray, R. and Wilder, J., "Robotic Acquisition of Jumbled Parts from Bins by Visual and Tactile Guidance," *Proc. 3rd Int. Conf. Robot Vision Sensory Controls*, Cambridge, MA, Nov. 1983, pp. 441–452.

Robertson, B. E. and Walkden, A. J., "Tactile Sensor System for Robotics," *Proc. 3rd Int. Conf. Robot Vision Sensory Controls*, Cambridge, MA, Nov. 1983, pp. 327–332.

Severwright, J. S., "Tactile Sensor Arrays–The Other Option," *Sensor Rev.*, Vol. 3, No. 1, 1983, pp. 27–29.

Severwright, J. S. and Baker, K. D., "Tactile Sensor Arrays for Flexible Assembly Automation," Colloquium on Robot Sensor Signal Processing, London, 1982, pp. 6/1-2.

Tanie, K., et al., "A High Resolution Tactile Sensor," *Proc. 4th Int. Conf. Robot Vision Sensory Controls*, London, Oct. 1984, pp. 251–260.

Trounov, A. N., "Application of Sensory Modules for Adaptive Robots," *Proc. 4th Int. Conf. Robot Vision Sensory Controls*, London, Oct. 1984, pp. 284–294.

Vranish, J. M., "Magnetoresistive Skin for Robots," *Proc. 4th Int. Conf. Robot Vision Sensory Controls*, London, Oct. 1984, pp. 269–284.

Williams, D. F., "A Tactile Sensing Method for Programming a Robot for Surface Following," *Proc. 6th Brit. Robot Assoc. Annu. Conf.*, Birmingham, UK, May 1983, pp. 183–192.

Other Sensors

Belforte, D. A., "Lasers and Robots," *Electro-Optics*, 1983, Vol. 15, No. 8, pp. 26–29.

Beni, G., Hornak, L. A., and Hackwood, S., "Proximity Sensing Uses Re-Entrant Loop Magnetic Effort," *Sensor Rev.*, Vol. 3, pp. 68–71.

Clergeot, H., Placko, D., and Monteil, F., "Imaging Using Eddy Current Sensors," *Proc. 3rd Int. Conf. Robot Vision Sensory Controls*, Cambridge, MA, Nov. 1983, pp. 349–356.

Derby, S. et al., "A Robot Safety and Collision Avoidance Controller," *Proc. 8th Int. Conf. Industrial Robots*, Detroit, June 1984, Vol. 2, pp. 21.33–21.43.

Dorf, R. C. and Nezamfar, "A Robotic Ultrasonic Sensor for Object Recognition," *Proc. 8th Int. Conf. Industrial Robots*, Detroit, June 1984, Vol. 2, pp. 21.44–21.60.

Grahn, A. R. and Astel, L., "Robotic Ultrasonic Force Sensor Arrays," *Proc. 8th Int. Conf. Industrial Robots*, 4–7, Detroit, June 1984, Vol. 2, pp. 21.1–21.18.

Hollingum, J., "Simple Solutions Trace Complex Surfaces," *Sensor Rev.*, 1983, Vol. 3, No. 2, pp. 81–83.

Hornak, L. A., Hackwood, S., and Beni, G., "Reentrant-Loop-Magnetic Effect Proximity Sensor for Robotics," *Proc. 3rd Int. Conf. Robot Vision Sensory Controls*, Cambridge, MA, Nov. 1983, pp. 357–362.

Klyueve, V. V., et al., "Robotic Sensors," *Prib. Sist. Upr.*, No. 1, 1983, pp. 15–17 (in Russian).

Knight, J. A. G., "Sensors for Industrial Robots: The State of the Art," *Proc. 2nd Eur. Conf. Automated Manufacturing*, Birmingham, UK, May 1983, pp. 127–132.

Koenigsberg, W. D., "Noncontact Distance Sensor Technology," *Proc. 3rd Int. Conf. Robot Vision Sensory Controls*, Cambridge, MA, Nov. 1983, pp. 371–384.

Luo, R. C., et al., "Object Recognition with Combined Tactile and Visual Information," *Proc. 4th Int. Conf. Robot Vision Sensory Controls*, London, Oct. 1984, pp. 183–196.

Marsh, K. A., "Acoustic Imaging in Robotics Using a Small Set of Transducers," *Proc. 4th Int. Conf. Robot Vision Sensory Controls*, London, Oct. 1984, pp. 261–268.

Meier, C., "Sensor Technology with the Robot Control M'," *Proc. 3rd Int. Conf. Robot Vision Sensory Controls*, Cambridge, MA, Nov. 1983, pp. 481–488.

Miller, G. L., Boie, R. A., and Sibilia, M. J., "Active Damping of Ultrasonic Transducers for Robotic Applications," *Proc. Int. Conf. Robotics*, Atlanta, March 1984, pp. 379–84.

Mitchell, I., Whitehead, D. G., and Pugh, A., "A Multi-Processor System for Sensory Robotic Assembly," *Sensor Rev.*, Vol. 3, No. 2, pp. 94–98.

Mortimer, J. M., "Leading the Way on Robot Weld," *Industrial Robot*, Vol. 10, No. 2, pp. 104–110.

Nitzan, D., et al., "Use of Sensors in Robot Systems," *Proc. Int. Conf. Advanced Robotics*, Sept. 1983, pp. 123–132.

Paul, R. P., "Sensors and the Off-Line Programming of Robots," *Proc. Int. Conf. Advanced Robotics*, Sept. 1983, Part 1, pp. 307–312.

Plander, I., "Intelligent Robot System with Vision, Tactile and Force Torque Subsystems," *Proc. 8th Int. Conf. Industrial Robots*, Detroit, June 1984, Vol. 2, pp. 19.120–19.131.

Zhao, C. J., et al., "Location of a Vehicle with a Laser Range Finder," *Proc. 3rd Int. Conf. Robot Vision Sensory Controls*, Cambridge, MA, Nov. 1983, pp. 409–414.

5

Robot Controllers

The controller is a major component of a robot which directs the
end effector to move in a desired sequence and to pass through desired points. It also functions to store position and sequence data
in memory so that the program can be repeated. Robot controllers
range in complexity from simple stepping switches through pneumatic logic sequencers, diode matrix boards, electronic sequencers,
and microprocessors to minicomputers. Simple data storage devices
(Table 5.1) in earlier robots used electronic counters, patch board
or diode matrices, and potentiometers. Other memory devices include magnetic tapes and discs, plated wires, solid-state semiconductors, and minicomputers with core memory. Program steps vary according to the memory device. Relay-type sequencers with printed
circuits offer 12 to 24 programming steps. Pinboard matrices provide up to about 300 steps, although steps of the order of 50 are
more common. Magnetic drum memories can provide up to 300 steps
and usually average about 200 steps. Operating systems in computer-based controllers may be hard-wired, stored in core memory,
or programmed in a read-only memory (ROM).

From the applications viewpoint, robot controls can be divided
into two major categories: nonservo control and servo control.
Servo control refers to the type of control in which the manipulator
motion is under constant supervision by a computer and requires
real-time feedback and analysis of the motion data. Nonservo control is much simpler and requires no feedback. In this chapter, we
will describe several types of controllers used in robots. A discussion of some typical systems will illustrate the practical aspects of
operation of robot controllers.

Table 5.1   Memory Devices in Selected Robots

Memory device	Source	Model
Mechanical step sequencer	Mack Corporation	BASE Series H
	Manca	Modular
	Modular Machine	Mobot
	Prab Robots	4200
	Seiko	100,200
	Sterling	Robotarm
Drum memory	Prab Robots	5800 HD
	Transcon	
Air logic	Auto-Place	10, 50
	Seiko	400/400L
Mechanical cam	Camel Robot	POI
	Pickomatic Systems	
Solid-state electronics	Accumatic Machinery	NuMan R1/R2
	Copperweld Robotics	CR-10, CR-50
	Cybotech	H80, V80
	Microbot	Minimover 5
	Pickomatic System	FR-200-4
	Systems Control	Smart-Arms
	Yaskawa Electric	Motoman

5.1   LIMITED SEQUENCE CONTROLLERS

The largest percentage of industrial robots are the pick-and-place type. Most of these simple robots are controlled at the basic level by hardware such as cams, ladder logic, sequential timers, and simple board-level microcomputers. Limited sequence robots are generally considered to be low-cost manufacturing tools with very simple kinematic, dynamic, and control systems.

The simplest robot control is found in mechanical pick-and-place robots typically powered by cam and follower mechanisms (Figure 5.1). In such robots the shape of the master cam determines the rotation and travel of the robot. A second cam rotating on the same shaft as the master cam usually controls the action of the gripper. The possible motions in this control are limited since the cams must be changed to alter robot movements. This type of robot control is useful in applications where high speed and great precision are required. The next level of control is obtained by using fixed stops on each mechanical joint or axis of the robot. These stops

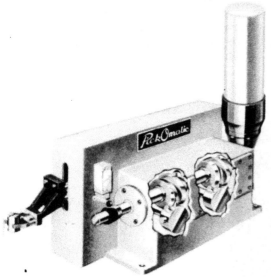

FIGURE 5.1   Roller gear cam-driven robotic parts handlers.
(Courtesy of Pickomatic Systems.)

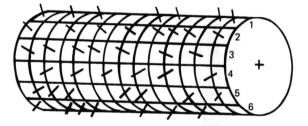

FIGURE 5.2   Mechanical drum controller.

may be adjustable so that the start and end positions of each joint
can be set for precise repeatability. Additional indexable stops,
which can be inserted or withdrawn automatically by a task program,
provide more than two-position control.

Some robots may also have simple fixed sequence controls which
actuate electric contacts by studded rotating drum and stepping
switches (Figures 5.2 and 5.3). Sequential timer methods are used
to control many simple pick-and-place robots. In a typical drum
timer, movable pegs on the drum open and close different solenoids.
This causes pneumatic or hydraulic fluid to actuate the various axes
of the robot. The travel of the axes is adjusted by mechanical
stops. The advantages of such a system are low cost of the control

FIGURE 5.3   Transcon drum controller.   (Courtesy of Transcon,
Inc.)

hardware and simplicity of programming. However, there is no provision for any conditional action, and the application must be well structured.

Programmable sequencers, such as printed-circuit boards, pneumatic logic modules, and pinboards, provide the capability for executing many motions in consecutive steps. Ladder or relay logic control requires a detailed description of the sequence of actions in terms of the opening and closing of relays. Programming of logic controllers involves knowing exactly how the relays must be wired together and how the logic components are to be attached to produce the required robot motion. Programmable controllers add more flexibility by permitting the control actions to be specified through a manual pendant or keyboard. Relay logic systems are inexpensive and provide enough capability for a simple, repetitive, and structured application. Pinboard-type robot control has openings arranged in rows and columns. Each row of openings represents one direction of motion for each axis, and each column represents a step in the program. In programming, a mutual setting of the pattern is effected by diode pins. The steps in the program can then be triggered by timing devices.

In typical pneumatic robots, the sequence of rotary and linear motions is preset by using a pneumatic logic circuit. At the start of program execution, the controller iniates signals to control valves on the actuators. The joints begin to move when the valves open and admit air to the actuators. Motion continues along an axis until the robot encounters a limit switch, which has been preset. The limit switch signals the end of travel to the controller, which commands the control valves to close. The sequencer then indexes to the next segment of the preset sequence. This may again be to the controller valves on the actuators on the gripper.

One widely used pneumatic system has modules for power and for sequencing. The double-acting pneumatic cylinders are actuated by an array of power, relay, and limit valves, plus other flow control devices and sensors. The flexible plastic tubing connecting these components is readily adjusted into a new pattern for a different robot sequence.

In a combination control using limit switches and mechanical end stops, the limit switches anticipate the positions of the stops and switch off the power stroke. Dampening begins to take effect and the arm coasts in against the mechanical stops, with resulting reliable and stable operation. Power may then be turned back on to hold the arm against the stop. This type of control provides comparative precision positioning as in cam-operated robots.

In a widely used robot of this type, internally mounted adjustable limit switches are synchronized with each mechanical stop. This provides an interface for closed-loop, feedback, and command

control circuits with peripheral equipment. In this way the movements of the robot can be coordinated in sequence with other operating equipment. For instance, the robot will not attempt to feed a
press until the press is in the open position, and the press will not
close until the robot has signaled its return to home position. Actuation of the robot joints is provided by a dedicated electromagnetic
air valve for each joint. The operational program is determined by
a controller, which programs the air valves in sequence and real-
time control and also properly manages the feedback circuitry. Typical controllers used may be (1) mechanical rotating drum-stepper
and (2) electrical relay and electronic programmable. Computer and
air logic may also be used.

Cycle speeds in nonservo robots tend to be much higher than
in servo-controlled robots because of the simpler control logic. The
jerky motion that results has given rise to the term "bang-bang" as
a descriptive reference to nonservo robots. Relatively high speeds
of operation are possible because of the smaller size of the manipulator and full air or oil flow through the control valves. Nonservo
robots are highly reliable and simple to operate and program. In
terms of program capacity and positioning capability, nonservo robots have limited flexibility. Repeatability to within ±0.25 mm
(0.010 in.) is achievable with smaller robots of this type.

The jerk motion in limited sequence or bang-bang robots produces damaging impact loads when the stops are hit. Parts may be
dropped and alignment of position may go out of control. To prevent this situation, energy-absorbing resilient devices may be installed on fast bang-bang robots (Figure 5.4). Two methods are
widely used. The first employs mechanical springs or resilient
bumpers. The second uses hydraulic or pneumatic dashpots in the
form of cylinder cushions and deceleration valves. Both devices
are nonlinear and produce high peak force loads during deceleration. Industrial hydraulic shock absorbers, with controlled linear
deceleration, are commercially available. A typical control system
for an electromechanical timer is shown in Figure 5.5.

In one pneumatically powered industrial robot, a microprocessor
controller operates the robot from memory and interfaces with other
equipment associated with the overall automated function. Programming is done on a hand-held teach module that uses simple graphic
symbols for robot commands. The manipulator base rotates to 180°,
adjustable in 15° increments. It has readily adjustable positive stop
adjustments on base rotate, reach, and lift axes, which use proximity switches. There is also an adjustable hydraulic dampening on
base rotate with proximity switch interlocks. The modules of the
robot controller feature plug-in solid-state relays with a lithium
battery-supported memory.

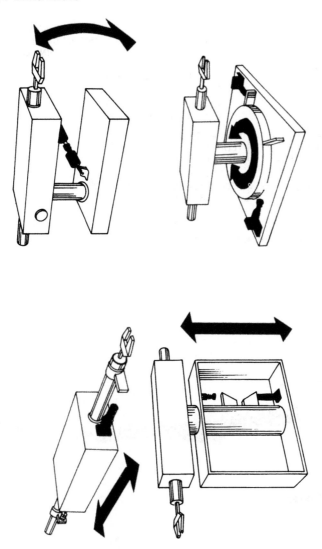

FIGURE 5.4   Resilient motion stoppers used in bang-bang robots.   (Courtesy of ACE Controls.)

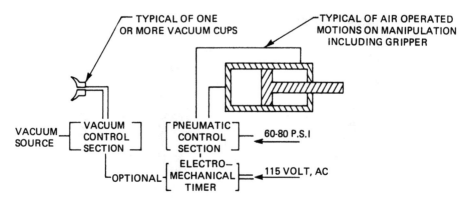

FIGURE 5.5   Typical control system for an electromechanical timer. (Courtesy of Mack Corporation.)

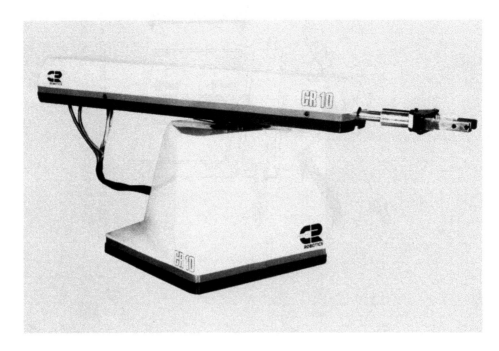

FIGURE 5.6   Pneumatically powered robots.   (Courtesy of Copperweld Robotics.)

The base rotation axis is driven by a rack-and-pinion system which is powered by dual air cylinders. The vertical lift and horizontal extend axes are also driven by air cylinder. A cam-actuated unit drives the wrist rotation, while the open/close grasp motion is driven by an air cylinder-actuated draw bar.

Models CR-10 and CR-50 of Copperweld Robotics (Figure 5.6) are pneumatically powered, microprocessor-controlled nonservo robots with proximity switches as the axis motion detectors. The CR-10, with a load capacity of 10 lbs, is used for a diversity of tasks which require parts handling, machine feeding, and assembly operations. The CR-50 is an exceptionally fast flexible device for handling workpieces weighing up to 25 lbs. Typical applications of the

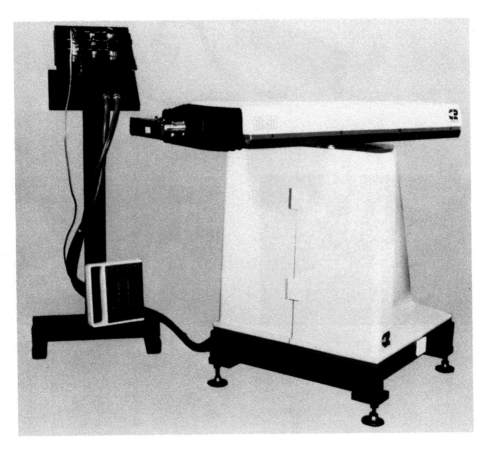

FIGURE 5.6   (Continued)

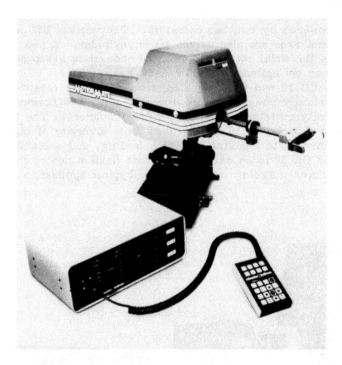

 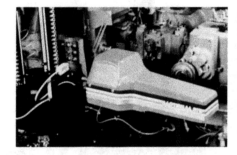

FIGURE 5.7   Motionmate robot system. (Courtesy of Schrader-
Bellows, a division of Scovil.)

CR-50 robot include machine loading and unloading, transfer of parts from conveyors, and tending injection molding and die casting processes. The control in each robot can communicate with as many as seven other machines.

Each axis of motion except the vertical lift has two adjustable fixed stops. The pneumatic (60 to 100 psi) axis drive system uses a pneumatic cylinder for the horizontal and vertical axes, an integral rotary actuator for the arm rotation axis, a rotary actuator for the arm turnover axis, and a cylinder actuator draw bar for the gripper.

Proximity switches are used for axis motion detectors and the electronic control is microprocessor-based. The nonvolatile EAROM has a memory capacity of up to 250 steps. The programming method is pushbutton via a portable teach pendant with full edit and jog modes. The time delays are selectable in increments of 0.1 sec from 0.1 to 99.9 sec. The program functions on the teach pendant include grip/release, extend/retract, up/down, over/clockwise/over counterclockwise, and rotate in/rotate out. See Figure 5.7 for another example of a robot with sequence control.

5.2   CONTROL WITH PROGRAMMABLE CONTROLLERS

Servo control in some electrical pick-and-place robots is typically provided by programmable controllers, using a microprocessor to perform logic skin to relay logic (Figure 5.8). Programmable controllers frequently have special software to perform many complex procedures according to simple instructions in user programs. The storing and accessing of data required for simulating the performance of electromechanical sequencers are facilitated by this software. Digital-to-analog (D/A) input and output modules are the means for effecting closed-loop control of robots by programmable controllers.

A typical programmable sequencer is equivalent to a rotating drum having several hundred sequential steps with a number of contacts at each step. A group of contacts in any one step indicates which robot axes to move during that step and in which direction. These data are then transferred to the programmable controller modules which command the required motions. Another set of contacts in a sequencer step shows the status of position-indicating devices at the end of the motion. The status is continuously monitored by the programmable controllers and compared to the corresponding sequencer contacts. The sequencer moves to the next step when all position indications match the sequencer step.

Alteration of contact arrangements is effected by changing the programs or by masking selectable contacts at each step. Robot

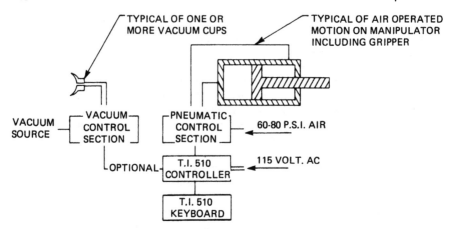

FIGURE 5.8   Typical control system for a solid-state programmer.
(Courtesy of Mack Corporation.)

movements can thus be varied by using different masks for different
robot tasks.

Sequencers are commonly programmed or loaded by means of a
teach mode.   In this technique the on-off contacts in the sequencer
are set to correspond to the on-off status of points.   A robot is
first jogged to a desired position.   Then the on-off status of all
pertinent inputs is copied into the sequencer step by pressing a
teach pushbutton.   The programmable controller is taught a sequence of movements by jogging the robot in this way through the
steps.

Closed-loop control of robots by programmable controllers is effected through D/A input/output modules, whose outputs are typically controllable over a range of -10 to $+10$ V DC.   The outputs
serve as a speed reference for the hydraulic or electrical servo
motor drive for each axis.   The servo motor shaft is mechanically
coupled to an encoder or potentiometer that feeds position and velocity data back to the programmable controller.   The processor de-
codes the encoder signal and uses this information to update and
modify motor velocity commands and acceleration rates.   Point-to-
point motion along each axis of the robot initiated and halted independently of the other axes is easily programmed on a programmable
controller.   However, vectored motion requires that the movements
along the axes be coordinated.   The programmable controller must
adjust acceleration rates and speed for each joint so that all motion
terminates simultaneously.   Vectored motion programs are more complex and require more memory than those for point-to-point motion.
Robot posit on calculations from encoder feedback can require

appreciable time. Besides, these data and those from external limit switches can be used for updating motor commands only once each scan. This puts a limitation on programmable controller servo control, which is feasible only if the errors resulting from the time delays are acceptable. Accuracy is also affected in long robot movements requiring great precision because of the resolution limitations of programmable controllers.

Typically, a programmable controller may support up to 32 registers that can be used, time registers, counters, and timers. It can perform matehmatical functions such as add, subtract, greater than, equal to, and not equal. Some programmable controller units use ladder logic programming and can be expanded to general I/O chassis. Random-access memory (RAM) in programmable controllers may be obtained with battery backup that can retain the memory for several months. EPROM is an unalterable nonvolatile memory that can be programmed on a program pendant and inserted into the programmable controller. Another feature of some controllers is the capability for off-line programming.

Stepping Motor Drives

Apart from servo motors, stepping motors may be used with programable controllers in robot control. A stepping motor shaft rotates through specific repeatable angular increments in response to an input pulse. Stepping motors are controlled by programmable controllers through stepping motor modules. These modules are typically microprocessor-based and store move instructions supplied by the processor in registers. On the run command from the programmable controller, the module converts programmed instructions into a pulse stream, which is fed to a stepping motor translator. This is an interface between a pulse source and a stepping motor. The translator amplifies pulses and translates them into the form required to produce stepping motor shaft rotation.

Stepping motor control is effected in either the open-loop or the closed-loop mode. In an open-loop system the module transmits pulses to the translator until the counter total equals a preset number. In closed-loop systems the stepping motor shaft drives an incremental encoder whose pulses are fed back to the module counter. The module produces pulses until the robot axis reaches the desired point. In closed-loop systems, encoders connected directly to stepping motor shafts produce errors caused by drive-gear backlash. Suitable coupling arrangements could offset these errors. In comparison to servo motors, stepping motors have limited acceleration and do not perform satisfactorily when driving high-inertia loads. Moreover, a stepping motor cannot be reversed rapidly because the module can store only one set of move instructions at a time.

FIGURE 5.9   Anthropomorphic robot driven by stepper motors or servo motors.   (Courtesy of Sigma Sales, Inc.)

The Sigma Max series robot shown in Figure 5.9 is driven by either stepping motors or servo motors.   The robot is computer-controlled and supports Fortran and Basic languages for robot control programs.   Max is a compact robot manipulator arm with six degrees of freedom plus hand open and close.   Optional parallel ASCII interfaces control the stepping rates of the motor.   A step-by-step interface that requires one character per step per channel is available.   An alternative vector interface requires one character sequence that defines the total motion of each channel, with execution time dependent on the component traveling the greatest distance.   Both of these interfaces are compatible with printer interfaces to minicomputer and microcomputer systems.   The vector interface uses a high-speed 16-bit microprocessor for real-time calculations.   The processor is controlled with an 8-bit control word, and data are communicated in 8-bit bytes with additional 1-bit Strobe

and 1-bit Ready. The interface calculates acceleration or deceleration ramp for each channel and executes moves as commanded by the
computer. It also senses the initial position of each axis and returns the status to the computer.

5.3   CONTROL WITH MICROCOMPUTERS

Small bench-top robots can be controlled by personal computers or
pocket calculators. These robots are typically used for experimental
and educational applications. Most of them have stepping motors
operated in an open-loop mode. The controller commands the actuators to rotate a given number of steps without any position feedback from encoders. The RS-232C interface is a common serial interface used in computer-controlled robots in which the bits of data
are transmitted one at a time, one after the other. It is usually
configured in a 25-pin connector known most commonly as the DD-25
connector.

The interface between a robot and a microcomputer often consists of an interface card that plugs into the microcomputer. This
card converts parallel data from the computer to serial data for the
robot and also handles transmit and ready handshake signals. The
design of the controller card may be based typically on an NMOS
chip such as the Intel 8748, which consists of CPU, EPROM, and
RAM. Commands to the arm joints are transmitted from the computer in the form of ASCII characters representing a specific operation. These operation code characters may be followed by numbers
which represent data for the operations.

In some robots parallel ASCII interfaces control the stepping
rates of stepper motor drives. Two interfaces are possible: (1) a
step-by-step interface that operates with one character per step per
channel and (2) a vector interface that requires one character sequence which defines the total motion of each channel. The vector
interface generally reduces the work load on the computer. The
keyboard of the host personal computer can also serve as a teach
pendant for a robot.

Many current commercial robots have microcomputers as programmable controllers. Microcomputers typically provide data reduction
by selective storage, programmable tracking accuracy, and task
modification during playback.

The microcomputer in such a robot is used to sample and record
positions in memory during the manual teach mode. Position recording can typically be done in a synchronous or asynchronous fashion.
In synchronous operation the sampling frequency is preprogrammed,
whereas in asynchronous operation the command to store a position
is given by pressing a button. In the playback mode the

microcomputer can be used to modify the task by sensory feedback signals. Another advantageous feature of a microcomputer is that it can be programmed to monitor the sensors and to indicate any failure in the system. For example, the microcomputer will terminate robot motion if an object is detected at an unexpected location.

A microcomputer can also be used as a preprocessor for a robot vision system to control the interface for digitizing the picture and storing it in memory. In this application the microcomputer will typically reduce the data and transfer the feature signature to a central computer. Another use of a microcomputer is as a tactile sensor processor to compare digitized sensor outputs with programmed threshholds and set appropriate status lines to the control computers.

Distributed microcomputer control of a robotic manipulator involves the use of more than one computer with coupled cooperating software. A manipulator is ideally suited for distributed control because it requires precise real-time coordination of the processors. A distributed processor architecture for a control system offers advantages such as functional and physical modularity. Functional modularity describes the ability of a system to remain constant over a range of system sizes. For example, the distributed control of a six-joint manipulator could be expanded to seven-joint control simply by adding one more controller. In the absence of distributed control, the entire system will have to be redesigned or reprogrammed to accommodate the expansion. High-volume, low-cost microprocessors offer a high degree of physical modularity.

A single processor used for all the actuators requires a complex time-slicing arrangement to service each actuator in sufficient time. This is subject to severe race conditions. For instance, at a time when the computer must monitor a particular actuator, it cannot relinquish time to maintain a current world model data base or communicate with an operator. The software to schedule the system to avoid such conditions tend to be very extensive. If separate processors are used for each servo in distributed control, the scheduling task can be conveniently allocated to another processor.

Criteria for distributed control of a manipulator are implemented in the following steps:

1. The servo action must be in joint space with trajectory end points specifying both position and velocity joint variables. This works reasonably even for tasks such as tracking a moving conveyor.
2. Approximate manipulator dynamic models may be used and each joint must be controlled optimally by a separate processor.
3. To make the system failure-tolerant, provision must be made for each processor to monitor the other processors and to have certain override capabilities.

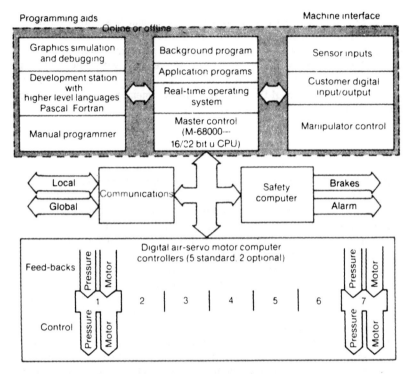

FIGURE 5.10 Computer and control hierarchy for a robot. (Courtesy of International Robomation/Intelligence.)

The robot distributed control shown in Figure 5.10 consists of a hierarchy of microcomputers with a Motorola MC 68000 16/32-bit processor as the master control computer. There is one Motorola 680X 8-bit microprocessor as the control computer for each of the six joints. A seventh 6800X microcomputer is employed as a safety control for monitoring all elements of the system. The system manager provides a real-time executive for handling both foreground and background tasks. It is segmented into four major areas, which are CPU, ROM (read-only memory), and I/O. The CPU has the advantage of 32-bit data addresss registers, 16-megabyte direct addressing range, memory mapped I/O, 14 addressing modes, and 56 instruction codes.

The CPUs of the axis controllers for distributed processing are configured with separate ROM, RAM, and I/O. The I/O sections primarily deal with the position encoder, pressure transducer, system communication, and diagnostic interface.

The robot may be programmed in hand commands, Macro Assembly, Fortran, or Pascal languages. Interfacing the robot with machine sensors is achieved by eight digital inputs and outputs. Each axis control computer shares 256 bytes of memory with the 68000 so that it can monitor the entire system. All data transfers for closed-loop control are made rapidly through direct memory address rather than as I/O transfers.

5.4   MINICOMPUTERS AND MICROPROCESSORS

Minicomputers and microprocessors play an important role in the control of more complex robot applications involving moving targets, assembly work, and batch manufacturing. For an example, see Table 5.2. Computer-aided teaching is mandatory in large manipulators with six degrees of freedom because of the impracticality of leading all joints individually for a specified end-effector trajectory. Teaching of individual points in a large multilevel parallel array of parts is also better done by a computer program.

In robots using medium- and high-level languages, a computer controller must be used. In one such configuration a 16-bit microprocessor-based minicomputer, LSI-11/2, is used. The LSI transmits commands to the manipulator joints via a joint interface, which in turn channels commands to the proper joint servos. The position commands from the LSI-11 are translated into motor control signals by the joint servos. Typically a joint servo consists of a dual-port memory communications buffer, a microprocessor-based controller, and a driver circuit. An interrupt logic may be used in place of the dual-port memory, but the latter is more efficient.

The components of the computer controller in Figure 5.11 are a teach pendant, LED status display, display selector switch, AUTO zero, test mode, reader control, CRT display, MDI keyboard, mode selector switch, task program load/record, setup panel, and EPROM memory control. The system has a basic RAM capacity of 16K for a total of 900 points. This can be reduced to 8K for simple applications or expanded to 64K for complex tasks. The programmable interface control (PIC) feature has a capacity for 4096 words of EPROM memory and performs the sequential control for machine logic. It provides an effective and simple way for the robot to communicate and interlock with other automated equipment such as machine tools and conveyors. For a typical computer control system, see Figure 5.12.

The program-edit feature enables an operator to make additions, deletions, and revisions to task programs. To edit a program, data in memory are called up on the CRT. Edit information is then entered character by character or in entire point-to-point blocks

TABLE 5.2   Robot Control System

Path control	Controlled path, closed-loop servo system.
Teaching	Remotely controlled using teaching pendant (included with robot) or CRT keyboard (full programs are entered at either). Optional off-line using absolute coordinates. Prompting and editing features are included, with ability to insert or change steps directly.
Teaching coordinate system	Cartesian (world, tool).
Logic	47 logic codes provide decision making to permit branching, various stop conditions, interrupts, external commands, arithmetic, etc.
External signal capability (inputs + outputs)	32 standard user available. All connections are optically isolated and available at a terminal block. Analog I/O available.
Program memory	Cassette tape or optional floppy disc.
Maximum number of points per program	2000
Maximum number of commands per program	2000
System executive functions	Error detection Automatic power fail recovery Diagnostic messages Program checking and identification Prompting controls Axis lockup

Source: Courtesy of MTS System Corporation.

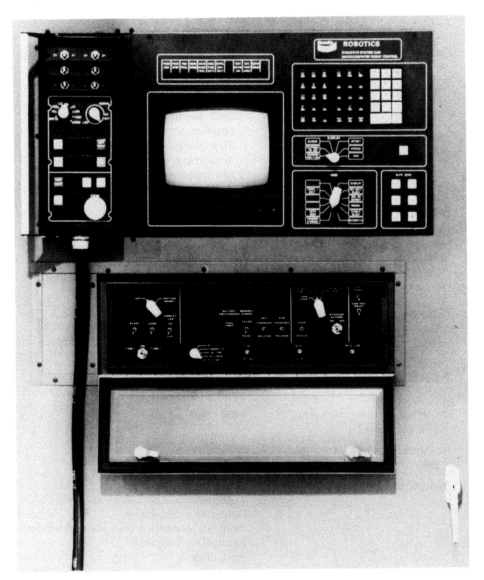

FIGURE 5.11   Dynapath system used as a controller for some
Bendix robots. (Courtesy of Bendix Robotics Division.)

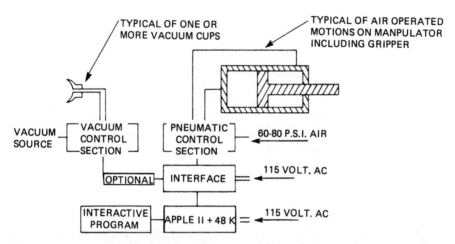

FIGURE 5.12   Typical computer control system for a robot.
(Courtesy of Mack Corporation.)

through the teach pendant. Program blocks are aligned, deleted,
added, and rearranged by means of editing features such as Step
Search, Block Search, and Step Edit. The LED status display indi-
cators on the operator's panel provide status and fault conditions
such as "overtravel" and "read error." The types of control inputs
supported are EIA (RS244) or ASCII (RS358) tape format, auto
block by block read-in, and test and diagnostic modes.

An example of a microprocessor-based positional controller is the
Anomatic II of Anorad Corporation. This controller uses microprocessor-based computer numerical control (CNC) with a built-in
calculator to realize an extremely high speed of operation. The
sealed touch pad keyboard (Figure 5.13) of the Anomatic II controller is separated into sections to identify the functions. The
data entry is similar to that of a calculator. Manual data input can
be done one command at a time or in memory for continuous operation. The jog function controls any of six joints in either slow or
fast single-step capability. It can be disabled via front panel
switch actuation. The calibrate/jog function is used to set the displayed values to those present before jogging. Return/jog is used
to move to the previous position before jog was invoked. Commands
in memory are displayed to facilitate editing and monitoring. The
typical status displays show position, status, offset, speed, sequence number, present command, next command, and messages to
the operator. Figure 5.14 shows a welding robot and its controller.

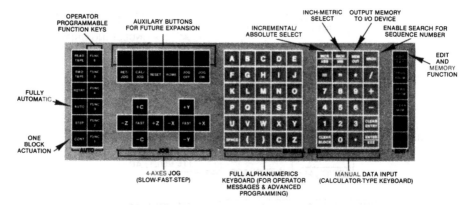

FIGURE 5.13   Keyboard and CRT display features of the Anorad
controller.   (Courtesy of Anorad Corporation.)

5.5   THE IRI M50 ROBOT CONTROLLER*

The IRI M50 robot has all of its control electronics on a single
"mono board."   It uses distributed processing with a hierarchy of
microcomputers and a Motorola 68000 16/32-bit microcomputer as the
system manager.   Each axis has its own Motorola 680X 8-bit microprocessor receiving and passing information to the manager.   A

---

*Reprinted by permission of International Robomation/Intelligence.

separate 680X microcomputer is dedicated as a safety control. The system manager provides a real-time executive for handling both foreground and background tasks. This allows users to program in IRI's Robot Command Language (RCL). A user with Computer-aided design/computer-aided manufacturing (CAD/CAM) can also link into the IRI M50 system. The manager provides central control over all axis computers. This high-level control architecture ensures that all axes are synchronized. The system manager or M 68000 is segmented into four major areas:

FIGURE 5.14   Welding robot and its controller.   (Courtesy of Advanced Robotics Corporation.)

CPU — The CPU is a Motorola 68000 16/32-bit microprocessor. This
    unit has the advantage of 32-bit data address registers, 16-
    megabyte direct addressing range, memory mapped I/O, 14 ad-
    dressing modes, and 56 instruction codes.
ROM — This is the location of the software that runs the system.
    The software developed by IRI to control the robot is RCL.
    The information in this section is fixed and can be changed
    only by changing proms (ROM).
RAM — All temporary or volatile information is stored in this section.
    This information includes the customer's command and point
    tables as well as the computer's scratch pad memory. A RAM
    capacity of 128K bytes is optionally available to the user.
I/O — This section connects the IRI M50 robot to the outside world.
    Through this section the unit receives its programming and in-
    structions from outside terminals, computers, or machines. It
    also uses this section to input or output status and command
    other machines. The I/O section includes two RS232 ports, the
    manual move pendant, and the option to interface 16 input and
    16 output lines (8/8 I/O standard). The axis controllers
    (680Xs), under distributed processing, control each axis. These
    CPUs are configured with separate ROM, RAM, and I/O. The
    ROM/RAM sections of the controller have about the same func-
    tion as they do in the system manager. They hold the oper-
    ating software needed to control the different modes of move-
    ment, including acceleration, velocity, braking, holding, deceler-
    ation, load adapting, and jogging. The I/O section of the axis
    controller is concerned primarily with the position encoder,
    pressure transducer, system communication, and diagnostic
    interface.

The position decoder interprets the incoming information from
the move encoders and tells the CPU direction and amount of move
made by the arm. The pressure transducers monitor the air pressure to ensure that enough pressure is available to do the work required. The diagnostic interface (a serial port) allows communication to the outside terminal. Information is transmitted to the system manager through the internal communication system.

The safety computer is also a 680X 8-bit microcomputer. It
monitors all elements of the system, and if they are not in agree-
ment it shuts down the system. The IRI M50 is shut down by in-
terrupting motor power, opening the 24-V supply line, and applying
the brakes.

See Figures 5.15 and 5.17 for illustrations of the IRI M50 robot
controller.

MONOBOARD BLOCK DIAGRAM

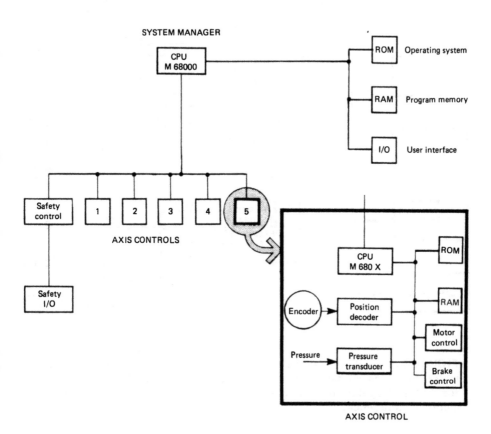

FIGURE 5.15   Single-board robot control unit with a six-layer
pointed-circuit board which performs 10 MIPS and features six 6800-
type axis processors running under control of one 6800 processor.
(Courtesy of International Robomation/Intelligence.)

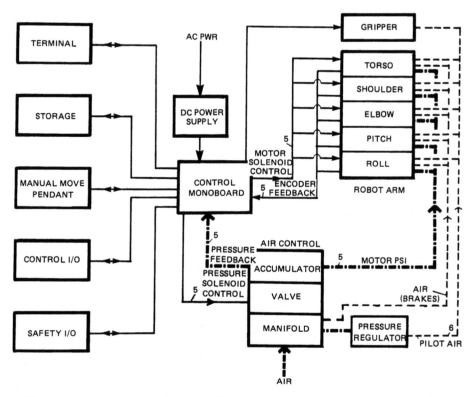

FIGURE 5.16   Block diagram of robot control having a monoboard which oversees all facets of robot operation and provides means for I/O and system programming.   (Courtesy of International Robomation/Intelligence.)

5.6   SPINE ROBOT CONTROLLER

The heart of the control cabinet is the processor system.   This is a distributed system with a master processor (MPA) and two slave processors (SPM).

The first slave processor, SPM/R, functions as a regulator in the control system.   The main purpose of the regulator is to steer the robot's seven axes so that the robot arm follows a required path or a path it has been taught.   The path is represented by different reference values in the form of space coordinates stored in the robot program.   In the second slave processor, SPM/T, the space coordinates are converted to robot signals.   The regulator transfers the robot signals (reference values) to the control signals for the

manipulator. SPM/T also converts signals from pendant and conveyor to space coordinates for storage or further processing in the
regulator.

The control system comprises valves, position sensors, power
sensors, and regulator. The posit on sensors are 11 potentiometers.
The power sensors sit between the hydraulic cylinders and the lines
and provide the control system with information about the tension in
the respective lines. The servo valves are controlled by signals
from the regulator SPM/R and distribute the pressure to the hydraulic cylinders. The signals which reach the regulator comprise *actual
values* from the sensors in the robot and *reference values* from the
master processor. All communication between the robot and the control cupboard is via the interface (CIN).

When the regulator receives an order from the master processor
to change the robot arm's position, a redistribution of the pressure

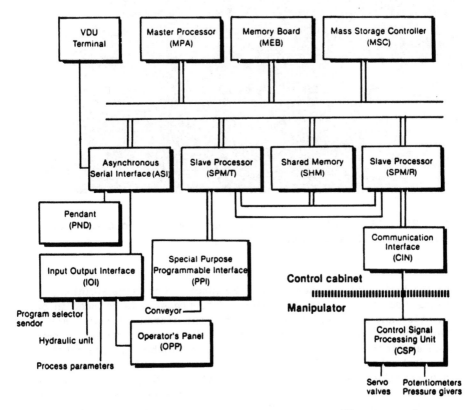

FIGURE 5.17   Control system for Spine robots.   (Courtesy of
Spine Robotics Corp.)

FIGURE 5.18   Control system hardware for Spine robots.   (Courtesy of Spine Robotics Corp.)

in the hydraulic cylinders must take place.   The redistribution takes place via the control system.   The pressure is reduced in one cylinder and increased in another, but the overall pressure, and thus the tension, is constant.   The tensile force decreases in one line and increases in another.   The result will be that the robot arm is bent in a particular direction and at a particular angle.   The process takes place in the same way but completely independant of the robot's under arm and upper arm.   The wrist is also controlled in the same way, but the hydraulic oil pressure operates directly on the wrist's three hydraulic motors.

   See Figures 5.17 and 5.18 for the Spine Robot Controller.

REVIEW EXERCISES

1. Explain the meaning of the following terminologies. Give examples and describe two or three industrial robots that use controllers of each type.

   (a) Servo and nonservo controllers
   (b) Limited sequence controllers
   (c) Programmable sequencers
   (d) Drum sequencers
   (e) Mechanical step sequencers
   (f) Electromechanical timers
   (g) Stepping motor control

2. Describe the features of robot microcomputer controllers.
3. What is distributed microcomputer control? Give examples.
4. Discuss the use of microprocessors and minicomputers in the control of industrial robots.
5. Sketch and explain the block diagram of the control system of any two industrial robots.

FURTHER READING

Barbera, A. J., Albus, J. S., and Fitzgerald, M. I., "Hierachical Control of Robots Using Microcomputer," *Proc. 9th Int. Symp. Industrial Robots,* Washington, DC, March 1979, pp. 405–422.

Bernorio, M., Bertoni, M., Dabbene, A., and Somalvico, M., "Programming an Industrial Robots in Italian," *Kybernetes,* Vol. 8, No. 4, 1979, pp. 305–313.

Bertino, M. and Furxhi, M. G., "Microcomputer Robot Control System," *Proc. 6th IECI Annu. Conf. Appl. Mini and Microcomputers,* Philadelphia, March 1980, pp. 367–372.

Cassinis, R., Schnickel, L., and Tomaini, M., "Economical and Powerful Microcomputer Based Stepping Motor Driver," *10th Int. Symp. Industrial Robots,* Milan, March 1980, pp. 89–100.

Coburn, J. D. and Emrich, K. B., "Controlling Robots with PCs," *Machine Des.,* Feb. 1982, pp. 55–60.

Dawson, B. L., "Moving Line Applications with a Computer Controlled Robot," *Robots II Conf.,* Detroit, Oct. 1977.

Hohn, R. E., "Application Flexibility of a Computer Controlled Industrial Robot," *1st North American Industrial Robot Conf.,* Rosemont, Oct. 1976.

Hohn, R. E., "Robot Control Systems and Applications," *Proc. Joint Automatic Control Conf.,* Denver, CO, June 1979, pp. 750–753.

Krut'ko, P. D. and Propov, Y. P., "Kinematic Algorithms for
    Manipulating Robot Movement Control," *Eng. Cybern.*, Vol. 17,
    No. 5, Sept. 1979, pp. 65–75.

Luh, J. Y. S. and Lin, C. S., "Multiprocessor Controllers for
    Mechanical Manipulators," *Proc. IEEE 3rd Int. Computer Soft-
    ware Appl. Conf.*, Chicago, Nov. 1979, pp. 458–463.

McCool, J., "Microprocessor in Control of Robots," *Electron Power*,
    Vol. 25, No. 11, Dec. 1979, pp. 796–799.

Park, W. T. and Burnett, D., "Interactive Incremental Compiler for
    More Productive Programming of Computer-Controlled Industry
    Robots and Flexible Automation Systems," *9th Int. Symp. Indus-
    trial Robots*, Washington, DC, March 1979, pp. 281–295.

Potter, R. D., "Practical Applications of a Limited Sequence Robot,"
    5th Int. Symp. Industrial Robots, Chicago, Sept. 1975.

Renard, M. and Iturrade, J. Z., "Robot Manipulator Control,"
    Proc. 9th Int. Symp. Industrial Robots, Washington, DC,
    March 1979, pp. 463–475.

Singer, A. and Rony, P., "Controlling Robots with Personal Com-
    puters," *Machine Des.*, Sept. 1982, pp. 78–82.

Skidmore, M. P., "Computer Techniques Used in Industrial Robots,"
    Industrial Robots, Vol. 6, No. 6, Dec. 1979, pp. 183–187.

Snyder, W. E., "Distributed Microcomputer Control of a Robotic
    Manipulator," *Proc. 1st Int. Symp. Mini and Microcomputer in
    Control*, San Diego, Jan. 1979, pp. 152–156.

Snyder, W. E. and Mian, M., "Microcomputer Control of Manipula-
    tors," *Proc. 9th Int. Symp. Industrial Robots*, Washington, DC,
    March 1979, pp. 423–436.

Tanner, W. R., "Basics of Robotics," Robot II Conf., Detroit, MI,
    SME Paper No. MS 77734, Oct. 1977.

Tanner, W. R., "Industrial Robots Today," *Machine Tool Blue Book*,
    Vol. 75, No. 3, March 1980, pp. 58–75.

Veira, J., "Programmable Controllers and Robotics—Which Is in Con-
    trol?" *Assembly Eng.*, Vol. 22, No. 10, Oct. 1979, pp. 22–24.

6
Nontextual Programming

6.1   PROGRAMMING WITH AN ELECTROMECHANICAL TIMER*

A good example of programming a very simple robotic system with an electromechanical timer is the ball transfer device shown in Figure 6.1A. This device transfers a steel ball from an open track (gravity-fed) to another location and repeats the cycle at 2-sec intervals. A sequence diagram (program) is shown as part of Figure 6.1A and consists of four translating motions with one gripper cycle during each sequence. This simple robot system uses three standard B.A.S.E robotic components to make up the manipulator, three air valves for the control section, and one timer for a programmer. The manipulator is assembled as shown in Figure 6.1B. Assembly weight is less than 5 lbs and the envelope is approximately 10 × 10 × 3 in. Mounting can be to any convenient structure and in any position. The control section consists of three 4-way, solenoid-actuated air valves, shown schematically in Figure 6.1C.

Figure 6.1D shows a typical electromechanical timer for use as the programmer. Timers have the reliability of an electric clock and are available off the shelf. The timer shown in Figure 6.1D has four microswitches with adjustable cams. All cams are located on a common shaft, which is driven by a synchronous motor through a

---

*Reprinted by permission of Mack Corporation.

PICKING A BALL FROM AN OPEN TRACK

GRIPPER MOUNTED IN LINE WITH FIRST MOTION

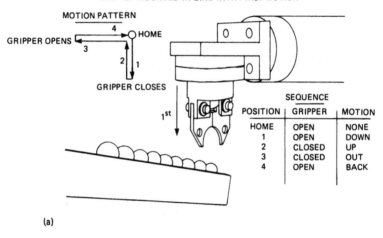

| | SEQUENCE | |
POSITION	GRIPPER	MOTION
HOME	OPEN	NONE
1	OPEN	DOWN
2	CLOSED	UP
3	CLOSED	OUT
4	OPEN	BACK

(a)

ROBOT FOR BALL PICKING TASK

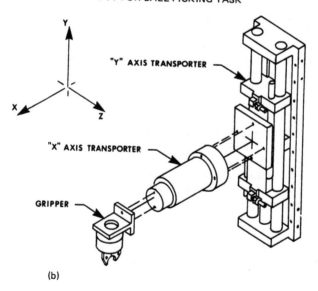

(b)

FIGURE 6.1   Programming with an electromechanical timer.
(Courtesy of Mach Corporation.)

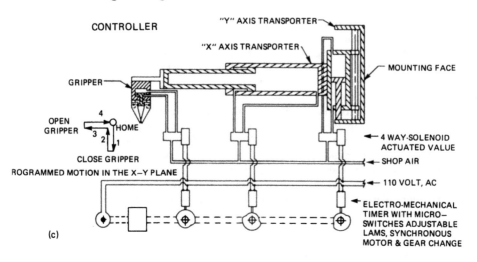

CONTROLLER

"Y" AXIS TRANSPORTER

"X" AXIS TRANSPORTER

MOUNTING FACE

GRIPPER

OPEN
GRIPPER

HOME

CLOSE GRIPPER

ROGRAMMED MOTION IN THE X–Y PLANE

4 WAY-SOLENOID
ACTUATED VALUE

SHOP AIR

110 VOLT, AC

ELECTRO-MECHANICAL
TIMER WITH MICRO–
SWITCHES ADJUSTABLE
LAMS, SYNCHRONOUS
MOTOR & GEAR CHANGE

(c)

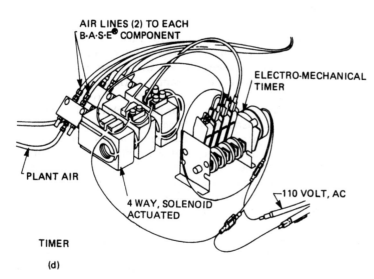

AIR LINES (2) TO EACH
B·A·S·E® COMPONENT

ELECTRO-MECHANICAL
TIMER

PLANT AIR

4 WAY, SOLENOID
ACTUATED

110 VOLT, AC

TIMER

(d)

FIGURE 6.1   (Continued)

choice of gear reductions. Cam shaft speeds of 1, 2, and 3 sec
provide a practical range. Setting up at the lowest speed is recommended. Fine tuning for higher speeds comes later. Cams are adjusted by trial and error with small end wrenches. One cam in each
pair sets up the time in which an event starts and the adjacent cam
controls the duration of this event. Microswitches have single-pole,
double-throw internal contacts. On each circuit where the solenoid
should be energized for less than 180° of cam rotation, use the normally closed switch terminal. On each circuit where the solenoid
should be energized for more than 180° of cam rotation, use the normally open terminal. Compensating for the "grief" in programming
electromechanical timers is simplicity and high reliability. This system and others like it have run 24 hours a day for millions of cycles
and never missed a stroke.

6.2 PROGRAMMING WITH A SOLID-STATE
PROGRAMMABLE CONTROLLER*

Programmable controllers provide a straightforward solution to on/off
commands and permit direct control over the start of an event as
well as its duration. Motion patterns shown in Figure 6.2 are typical of many which can be programmed directly with controllers.
Further, most controllers have sufficient memory to program all three
patterns (plus more) in one continuing "master cycle." The manipulator shown in Figure 6.2 is assembled from five B.A.S.E robotic
components consisting of a gripper, three transporters, and one intermediate stop cylinder. The stop cylinder provides a precision
midpoint stop location along the Z axis when called for in a program.
A system schematic of Figure 6.2 is shown in Figure 6.3B, where
five robotic components are plumbed to shop air and hardwired to a
programmable controller. A typical motion pattern for this robot is
shown in Figure 6.3A.
      This motion sequence could be called "pick-dip-drop." An ob-
ject is picked up at station 2, dipped or otherwise processed at station 6, and dropped at station 9. In programming a B.A.S.E. robot,
the first step should be to determine a "home position." The home
position is arbitrary, but it usually represents the gripper location
at the start and end of a sequence. The status of home position is
shown in Table 6.1A. Presetting a B.A.S.E. robot to home position
is accomplished easily by turning power off to the solenoid valves
and turning air on to the system. Reverse air lines to any component which is not in the home position. Presetting a home position

---

*Reprinted by permission of Mack Corporation.

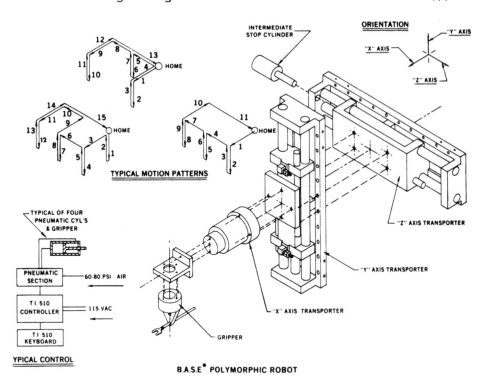

FIGURE 6.2 Modular Robot (courtesy of Mack Corporation).

with solenoids in the off mode is important in that all robot motions are now referenced to on/off commands. For example, the X axis transporter is retracted in the home position with an off command; therefore an on command will extend it. Presetting to the home position with power off has another advantage. In case of power failure, the robot drops everything, returns to the home position, and stays there unless programmed otherwise. On resumption of power, B.A.S.E. robots can be programmed to start from the beginning, return to the interrupted motion, or halt all movements until the entire system is reset by an attendant.

An abbreviated summary of motions related to on/off commands is shown in Table 6.1B. To program the sequence in Figure 6.3A plan each motion and record the status of each command without regard to time.

The program plan in Table 6.1C can now be programmed on a solid-state controller. For this example, we are using a Texas Instrument Model 510, shown in Figure 6.3C. Programming is easy

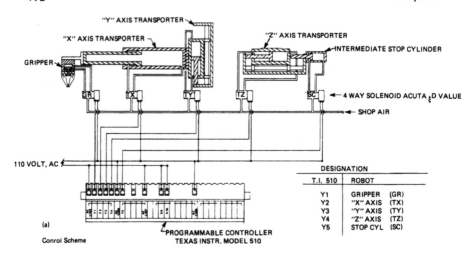

T.I. 510	ROBOT	
Y1	GRIPPER	(GR)
Y2	"X" AXIS	(TX)
Y3	"Y" AXIS	(TY)
Y4	"Z" AXIS	(TZ)
Y5	STOP CYL	(SC)

(a)
Conrol Scheme

PROGRAMMABLE CONTROLLER
TEXAS INSTR. MODEL 510

MOTION SEQUENCE OF
PICK–DIP–DROP TASK

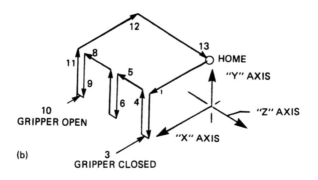

(b)

TYPICAL PROGRAMMABLE CONTROLLER

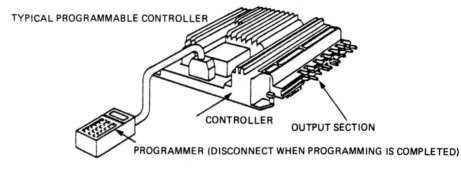

CONTROLLER   OUTPUT SECTION

PROGRAMMER (DISCONNECT WHEN PROGRAMMING IS COMPLETED)

(c)

FIGURE 6.3   Programming with a solid-state programmable controller.
(Courtesy of Mack Corporation.)

TABLE 6.1 Programming with a Solid-State Programmable Controller

A. Home Position

Gripper	Open
"X" Transporter	Retracted
"Y" Transporter	Up
"Z" Transporter	Right
Stop cylinder	Retracted

B. Conditions vs. Commands

Valve Position	Condition	Component	Condition	Valve Position
Off	Open	Gripper	Closed	On
Off	Retracted	"X" Transporter	Extended	On
Off	Up	"Y" Transporter	Down	On
Off	Right	"Z" Transporter	Left	On
Off	Retracted	Stop cylinder	Extended	On

C. Motion Sequence

		Command Status				
Step	Motion	GR	TX	TY	TZ	SC
1	TX-Out	OFF	ON	OFF	OFF	OFF
2	TY-Down	OFF	ON	ON	OFF	OFF
3	GR-Close	ON	ON	ON	OFF	OFF
4	TY-Up	ON	ON	OFF	OFF	ON
5	Tz-Left	ON	ON	OFF	ON	ON
6	Ty-Down	ON	ON	ON	ON	ON
7	TY-Up	ON	ON	OFF	ON	ON
8	TZ-Left	ON	ON	OFF	ON	OFF
9	TY-Down	ON	ON	ON	ON	OFF
10	GR-Open	OFF	ON	ON	ON	OFF

TABLE 6.1   (Continued)

11	TY-Up	OFF	ON	OFF	ON	OFF
12	TX-In	OFF	OFF	OFF	ON	OFF
13	TZ-Right	OFF	OFF	OFF	OFF	OFF

D.   Drum Timer Programming Form

Drum NR =   Preset =   Scan/Count =

Step	Counts Per Step	Outputs					
		Y1 Gr	Y2 Yx	Y3 Ty	Y4 Tz	Y5 Sc	
1	60	OFF	ON	OFF	OFF	OFF	
2	60	OFF	ON	ON	OFF	OFF	
3	60	ON	ON	ON	OFF	OFF	
4	60	ON	ON	OFF	OFF	ON	
5	60	ON	ON	OFF	ON	ON	
6	60	ON	ON	OFF	ON	ON	
7	60	ON	ON	OFF	ON	ON	
8	60	ON	ON	OFF	ON	OFF	
9	60	ON	ON	ON	ON	OFF	
10	60	OFF	ON	ON	ON	OFF	
11	60	OFF	ON	OFF	ON	OFF	
12	60	OFF	OFF	OFF	ON	OFF	
13	60	OFF	OFF	OFF	OFF	OFF	HOME
14	0						
15	0						
16	0						

Source: Courtesy of Mack Corporation.

and consists of loading an electronic drum using the following
keystrokes:

```
STR  X9
STR  X10
AND  NOT  C1
DRM  #1 (See Table 6.1D)
OUT  C1
END
```

A brief description of the drum programming form follows.
Drum number 1 was selected and only a small portion was used.
There are four drums available for independent use or in sequence
for much more comprehensive programs. Preset entry indicates the
number of the step to which the drum returns when reset. Scan /
Count is the number of 16.67-msec periods in each count. Step is
taken from Table 6.1C. There are 13 steps to this program, and
the drum can handle up through 16 steps. Counts-per-step entries
represent the number of counts that elapse before the drum advances to the next step. The time spend in a given step is equal
to the number of counts per step multiplied by the scans per count
times 16.67 msec. A zero entry causes that step to be skipped.
For our program, we have selected a number which provides a slow
stepping rate of approximately one step per second. One step per
second permits verification of motions and represents good setup
technique. Decreasing cycle time can be added later with minor re-
finements to the program. The cycle time for a 13-step program
such as this can be reduced to less than 5 sec. In the output columns the designation in the upper left-hand portion of each column
identifies the Texas Instrument output terminal. The letters in the
lower right-hand portion indicate which B.A.S.E. robotic component
is served. On /off commands are entered through simple digital conversion to 0's and 1's.

Typically, five basic ladder logic instructions can be used to enter and combine all ladder logic elements in a program. These instructions are: string or store (STR), output (OUT) series (AND),
parallel (OR), and reverse action (NOT). The STR instruction is
used to begin each new line (or group) of ladder logic. The OUT
instruction is used to create an output element in ladder logic. It
designates the last element in a ladder logic line. The elements preceding it in the line drive this output element. The AND instruction
is used to combine ladder logic input /output (I /O) elements in series.
The OR instruction is used to combine ladder logic I /O elements in
parallel. Finally, the NOT operation is used to reverse the normal
operation of a ladder logic I /O element. Series-parallel inputs require a combination of the basic instructions such as AND-STR

and OR-STR. AND-STR combines the last groups of elements be-
ginning with STR in series with the preceding group beginning with
STR in parallel with the preceding group beginning with STR.
Specialized functions such as latched relay may also be used.

6.3   LEAD-THROUGH PROGRAMMING

In lead-through programming, the operator leads the robot through
the desired positions by means of a remote teaching pendant. This
type of programming system is one of the most common on industrial
robots. Depending on the manufacturer, the hand programming unit
may have motion pushbuttons and joysticks as well as command keys
to set the speed, time delays, and external outputs and inputs.
     A task is taught, typically, by setting the correct switches on
the control panel. The manipulator is moved by means of the joint
switches and the required positions are recorded. Task parameters
such as the gripper state, time delays, and travel speed may also
be set and recorded. On playback in the automatic mode an appropriate control algorithm is used to generate the path between the
taught positions. The possible paths include straight-line motion,
circular arc interpolation, joint interpolated motion, and continuous
path motion.
     In some robots a point-to-point teaching mode is accomplished
by a manual module that has a rate-control switch for each joint.
Such robots have solid-state electronically controlled servo mechanisms, a programmable read-only memory, and a dedicated minicomputer for robot action control. The switches on the manual teaching
box allow the operator to guide the robot to the desired position.
When a record button is pushed, the position of each joint is stored
in memory. On playback of the taught program, there is point-to-
point servo operation for all joints simultaneously in the most efficient path.
     The software structure in teach pendant programming is inter-
pretive and records only the end points. The interpreter has the
capability to execute special library routines for some logical branching and other functions. In addition, the interpreter allows execution of the program one step at a time. Editing features such as
step deletion, modification, and insertion are easy to implement, because library routines provide instantaneous position information to
the servos during playback. Practical features are as follows:

Advantages

1. Many special functions such as time delay, output signal rates,
    and travel speed can be programmed.
2. Editing capability is better than in walk-through programming.

FIGURE 6.4   Robot hand programming unit.   (Courtesy of Expert
Automation, Inc.)

FIGURE 6.5   Teach pendant.   (Courtesy of Expert Automation, Inc.)

3. Longer programs can be handled because of the discrete end-point positions.

Disadvantages:

1. Simple programs with limited branching.
2. Limited provision for documentation.
3. Only on-line programming is possible.

The hand programming unit shown in Figure 6.4 has a pair of keys for the right-hand and left-hand motion of each axis.   The axes of the robot are moved by touching the appropriate keys until the end effector reaches the desired location.   These final points are stored in memory for the automatic motion during program execution.   The display units indicate functions and program steps and can also be used for troubleshooting.

The hand programming unit in Figure 6.5 is used to input movement paths, stopping points, and speeds into the control system. Also shown in the figure is the drilling of holes in curved metal sheets.   The two drilling units are used alternately and are kept constantly perpendicular to the complex surface of a workpiece.
The operating elements of the hand programming unit are subdivided into four function blocks.   These are two line displays with 16 characters each.   LED displays for the operating status, manual mode

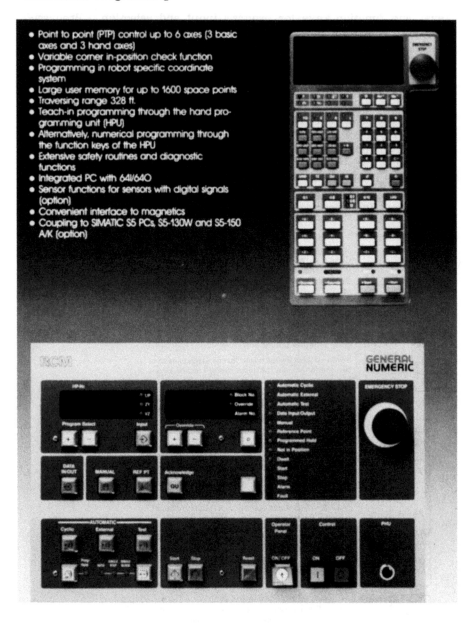

- Point to point (PTP) control up to 6 axes (3 basic axes and 3 hand axes)
- Variable corner in-position check function
- Programming in robot specific coordinate system
- Large user memory for up to 1600 space points
- Traversing range 328 ft.
- Teach-in programming through the hand programming unit (HPU)
- Alternatively, numerical programming through the function keys of the HPU
- Extensive safety routines and diagnostic functions
- Integrated PC with 64I/64O
- Sensor functions for sensors with digital signals (option)
- Convenient interface to magnetics
- Coupling to SIMATIC S5 PCs, S5-130W and S5-150 A/K (option)

FIGURE 6.6   Features of a teach pendant.   (Courtesy of General Numeric.)

keys, and function keys for program input and selection. Programming with the teach pendant is possible only in manual mode operation.

Figure 6.6 shows the General Numeric RCM-1 point-to-point robot control system. Programming of the control is simplified through a programming language that is based on symbolic abbreviations and through a programming structure with subroutines. The hand programming unit has the operating elements subdivided into function blocks which include manual mode keys and function keys for program input and selection. The operator panel is subdivided into groups such as numerical displays, mode of operation keys, and program selection. The logic unit is equipped with several interfaces to the hand programming unit, magnetic tape cassette recorder, and incremental position encoders. The user program is

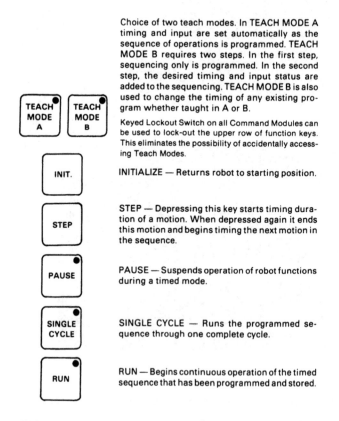

Choice of two teach modes. In TEACH MODE A timing and input are set automatically as the sequence of operations is programmed. TEACH MODE B requires two steps. In the first step, sequencing only is programmed. In the second step, the desired timing and input status are added to the sequencing. TEACH MODE B is also used to change the timing of any existing program whether taught in A or B.

Keyed Lockout Switch on all Command Modules can be used to lock-out the upper row of function keys. This eliminates the possibility of accidentally accessing Teach Modes.

INITIALIZE — Returns robot to starting position.

STEP — Depressing this key starts timing duration of a motion. When depressed again it ends this motion and begins timing the next motion in the sequence.

PAUSE — Suspends operation of robot functions during a timed mode.

SINGLE CYCLE — Runs the programmed sequence through one complete cycle.

RUN — Begins continuous operation of the timed sequence that has been programmed and stored.

FIGURE 6.7   Teach modes of MotionMate. (Courtesy of Schrader-Bellows, a division of Scovil.)

FIGURE 6.8   Teach pendant.   (Courtesy of Advanced Robotics
Corp.)

modular segmented in comprehensible sections with a total of 99
programs having 99 subroutines, branches, and cycle programs.
Program control instructions are jump, conditional, swell time, hold
stop, and cancel remaining runs. Movement instructions are movement block, velocity selection, and corner in position check. Load
parameters, add, subtract, multiply, and compare parameters are
the arithmetic instructions. The model RCM-2 controller is designed for continuous path control and has more control features
than the RCM-1. Features of the control include continuous path
with three-dimensional linear interpolation, programming in Cartesian
coordinate system or joint coordinates, arithmetic functions programmable at the operator panel, and conveyor synchronization. See
Figures 6.7 and 6.8 for other examples. Programming Instructions
are listed in Table 6.2.

TABLE 6.2   Programming
Instructions

Program control instructions

   Jump instruction
   Conditional instruction
   Dwell time
   Hold for input
   Programmed stop
   Cancel remaining runs

Movement instructions

   Movement block (PTP)
   Velocity selection
   Corner in position check

Input-output instructions

   Set, reset
   Gripper open, close
   Auxiliary function output
   Parameter output
   And /or function

Arithmetic instructions

   Load parameters
   Add, subtract, multiply
   Compare parameters

Special instructions

   Text output
   No operation

Source: Courtesy of General
Numeric.

FIGURE 6.9   Lead-through programming.   (Courtesy of ESAB
Robotic Welding.)

TABLE 6.3   Sample Interactive Input File
for **ORTHOCENTER** Simulation Program

---

input the information as asked for

R1 — floor to waist?   20
R2 — offset at waist?   5
r3 — length of upper arm?   15
R4 — offset at elbow?   4
R5 — length of forearm?   15

now we need to input information about the
angles

max. rotation of waist?   180
min. rotation of waist?   −180
max. angle of waist?   180
min. angle of waist?   −180
max. angle of elbow?   180
min. angle of elbow?   −180
value of motion index?   5
initial value of the a?   0
initial value of phi?   0
initial value of psi?   0
location of box center x,y,z?   10,10,10
box size 10

Make sure number lock is pressed

7-theta —   8-theta +
4-phi —   5-phi +
1-psi —   2-psi +

t or T — go to top view

s or S — go to side view

esc is used to end the program

r or R is used to restart

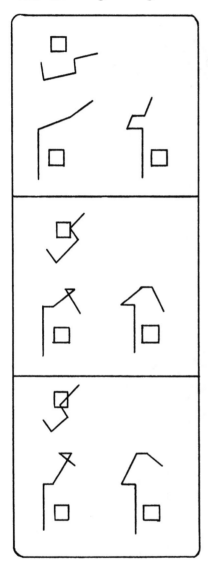

TABLE 6.10   Simulation of a robot in a workspace with cubic obstacle.

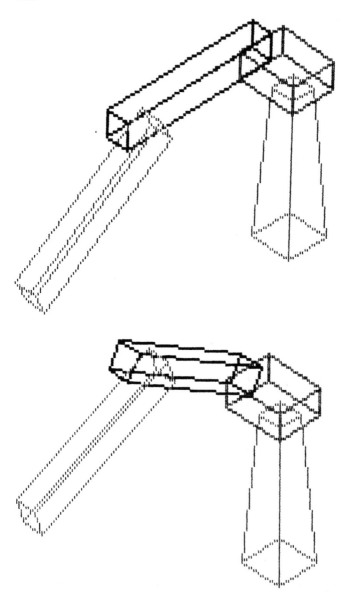

FIGURE 6.11   Simulation of robot movement with TDROBOT.

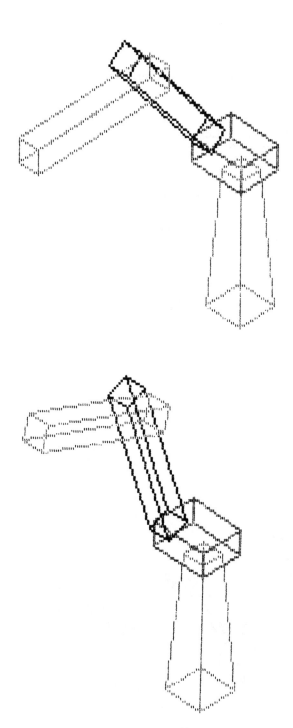

FIGURE 6.11 (Continued)

TABLE 6.4   Use of Numeric Keyboard in Centerline and
Three-Dimensional Simulation Program

PUMA robot in the zero position. This structure can either
be changed to a different configuration or a 3D structure
can be created around the stick figure.

(NUM LOCK SHOULD BE ON)

The program is controlled by the following keys:

'7' — DECREASE THETA 1   '9' — INCREASE THETA 1
'4' — DECREASE THETA 2   '6' — INCREASE THETA 2
'1' — DECREASE THETA 3   '3' — INCREASE THETA 3

'0' — ZERO POSITION
'5' — 90 DEGREE POSITION

'1' — TO CHANGE ANGLE INCREMENT (DEFAULT = 10 DEG)

HIT ANY KEY TO CONTINUE

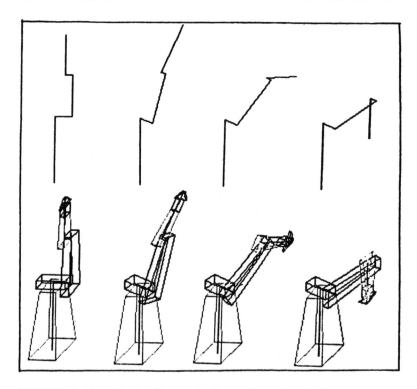

FIGURE 6.12   Centerline and three-dimensional figures.

6.4   LEAD-THROUGH TEACHING IN
SIMULATION

In this section we will briefly describe three computer programs
used for the simulation of robots. These programs are designed to
move the schematic of a robot on the screen in a lead-through teaching mode. The keyboard is converted to a teach pendant for moving
each joint of the robot. See Figure 6.9 for an example.

The first program, QRTHOCENTER, displays orthogonal projections of the movements of an anthropomophic robot. A centerline
or stick figure is used throughout. This program also has a cubic
obstacle whose location and size are input interactively by a user.
Configuration data, such as values of ϕ, θ, ψ, may be shown at
any point in the program as desired by means of a toggle key.
Table 6.3 shows the interactive input and typical results are shown
in Figure 6.10.

The second program, TDROBOT, is developed along the same
lines as ORTHOCENTER. However, the graphical display uses a
three-dimensional solid model fo the robot (Figure 6.11). In both
programs the command CMD$ = INKEY$ is used to read the keyboard
buffer, thus eliminating the need to press the enter key during the
control loop.

In the COMBINER program, both centerline diagrams and three-
dimensional representations are used. The centerline diagram is
used in driving the robot to a desired point. At any such point
the complete three-dimensional diagram is obtained by pressing the
space bar. The way in which the numeric keypad is utilized is
shown in Table 6.4. Typical results are presented in Figure 6.12.

6.5   WALK-THROUGH PROGRAMMING

Walk-through programming is a common technique for robots used in
finishing applications. To teach the robot a task, an operator manually walks the robot through the required motions by means of a
teaching aid. Two walk-through teaching aids normally employed
are (1) special attachments for the manipulator wrist and (2) a
special teaching arm. In this teaching process, switches and pushbuttons on the control panel are used for specifying program number and sampling rate for recording points in a mass storage device.
Other functions which may be set on the control panel are remote
on/off and a synchronization program for conveyor tracking.

The programming module of a typical walk-through robot holds
a tape deck and a standard frame containing electronic circuit cards.
When this type of robot is to be programmed, a switch on the console is set to "teach." Typically, an operator physically grasps the

terminal device and guides it through the desired path and sequence
in the exact manner and speed at which the robot is to repeat the
motion. Signals generated by the motions are transmitted in analog
form, then converted to dig tal language in the control console. A
trigger on the terminal device permits the speed and direction of
the motion pattern to be recorded. While the device is moved
through the desired path, the posit on of each joint is recorded on
a constant time basis, thus generating a continuous time history of
each joint position. If the operator makes a mistake during the
walk-through, a trigger is released allowing the control tape to reverse at high speed. At the desired moment the trigger is compressed again to start the correct sequence. New signals automatically erase the previous ones on the tape.

Since the operator must grasp the robot in programming, it must
be designed to be essentially light and free. Some robots are constructed in such a way that the arm is in a state of continual equi-
librium so that it can be manually moved accurately, smoothly, and
substantially inertia free. In other designs, the hydraulic cylinders
which power the manipulator may be disconnected from the robot
arm so that they do not offer any resistance to the movement of the
arm during programming. In some spray-painting robots, walk-
through programming is effected by manually moving a separate
articulated programming arm through the desired spray application
pattern. This lighweight independent component is counterbalanced
and permits the programmer to point with natural and normal
movement.

The software organization for walk-through programming basically consists of an interpreter which samples and records position and
functional on/off data into mass storage during teach mode. The
major features of this type of programming are summarized below:

Advantages:

1.   Ease of programming
2.   No knowledge of robot language is necessary.

Disadvantages:

1.   Lack of good editing capability.
2.   Reprogramming is necessary for correction.

The robot system shown in Figure 6.13 consists of a six-axis
manipulator, control console, and hydraulic power supply. The
manipulator is a hydraulically driven servo-controlled robot arm. It
is programmed for spraying by the walk-through method. Quick-
disconnect programming pins located at the lower arm (axis 2) and
the upper arm (axis 3) can be removed during the teach process

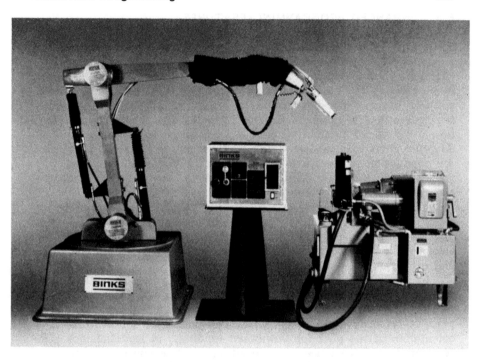

FIGURE 6.13   Components of a walk-through robot. (Courtesy of
Binks Corporation.)

for easier manipulator movement. To move the manipulator easily
during programming, bypass valves are located on the manipulator.
In the open position these valves allow hydraulic flow freely around
the actuators.

The manipulator arm is manually moved through a desired spray
sequence by gripping the two handles. The sequence is recorded
by continually depressing the program trigger. To record a point-
to-point program, the switch may be turned on and off at each of
the desired points. A microcomputer-based solid-state control system with closed-loop servo and analog feedback is used on this
robot.

The expandable memory size is 4.2 minutes continuous path and
the number of programs is eight with expandable capabilities. The
interface hardware is a terminal strip for external wiring.

6.6   ENHANCED PROGRAMMING FEATURES

After a program has been taught, there may be a need to make
alterations in some places.   Some robots have an editing system
which accomplishes this.   The editing system allows small changes
to be made in the program during each operating cycle until the
optimum program has been achieved.   Such an editing system allows
the corrections to be made without removing the robot from production.   For added flexibility, some walk-through robots have an
adaptive correction system which continuously adjusts the program
for changes in temperature, hydraulic fluid viscosity, and load.

In application of continuous-path robots, an operator teaching
the robot motions may also program the tool functions along the
path.   For example, some spray-painting robots permit programmable spray gun functions.   To provide greater flexibility the operator may program a spray gun flipper, a spray gun pattern changer,
electrostatic spray gun voltage on/off, activation of color changers,
and other job-related auxiliary functions.

In programming welding robots for continuous-path motion, it
may be necessary to alternately jog the part and the welding torch
into position for welding along the desired relative path.   The workpiece is typically mounted on a two-axis positioner for continuous
orientation of the part for downward welding.   A lead-tailstock combination may also be used as a positioner.   Numbered reference
marks are generally made at the end points of the programming
blocks.   The robot is then programmed to establish either the line
or the velocity between these marks.   The marks serve to determine
the relative motion between torch and part during the three-dimensional manipulation in the welding process.

An automatic reteach system, with a detachable tactile probe,
enhances the robot's ability for complex weldments.   The tactile
probe helps the robot to re-edit its own program from a rough
taught path.   It may also be used to re-edit a program where dimensional variation from part to part is outside allowable limits.
Other programming assists such as floating zero offsets, symmetry
modes, and circular, helical, and elliptical interpolation are useful
for programming multiple robots.   Use of multiple robots programmed
together reduces part distortion through uniform heat.   Some robots
may be programmed to weld at corners continuously without cutting
the arc while the robot computes its track.

Conveyor speed synchronization is an enhanced feature in programming robots for flexible and efficient production activities.
Synchronization may be achieved by encoder feedback from the conveyor to compensate for conveyor speed changes, including starts
and stops.

The enhanced features of spray-painting robots provide a good indication of the versatility of a robot. Some of these features are linear-controlled search function, jump and waiting functions, test-wait and repeat functions, and pattern and grip functions. Linear control permits the programming of curves at varying speeds. In this case the curve is divided into a number of straight-line sections selected on the basis of the required accuracy. Picking objects from racks of varying heights and searching for corners are facilitated by the inear-controlled search function. The jump function within a program or to another program permits simultaneous tending of machines that are not synchronized. Times between 0.1 and 99 sec can be selected by the waiting function. In test-wait function mode a robot stops and waits until it receives a signal via an input before it continues executing the program. The pattern function simplifies programming for pattern picking and placing.

An overview of the enhanced programming features of an electric arc welding robot is obtained by summarizing its simplified teaching and tracing functions, special welding functions and programming.

Simplified teaching and training functions:

Linear interpolation between start and end points
Circular interpolation for three end-points
Jog teach function for precision incremental position during programming
Operation in either absolute Cartesian coordinate system or relative Cartesian coordinate system
Multiple simultaneous axis positioning during programming
Forward and backward tracing
Input/output display panel for confirming the program and checking external output and welding output

Special features for welding:

Wrist interpolation for a variety of bend axis positions
Wire inching forward and/or backward for precise set of the welding operation
Weaving with variable frequency, pitch, and pause time
Built-in welding sequences
Programmable welding parameters such as wire feed and voltage
Continue function to restart welding from a halt command position
Welding parameter override for changing wire feed rate and voltage in three increments without disturbing the program

Programming and editing:

   Speed edit function enables speed of path segments to be
      changed individually or in groups
   Program jump function allows a program to be skipped either
      directly or as a result of a conditional external input

6.7   PROGRAMMING FEATURES OF SPINE
         ROBOTS*

The pendant is used for teaching, programming, and editing and
for manual operation of the robot. It is connected to the control
cabinet with a cable. The pendant contains indicator lamps, a
character window, and a pad with different function keys. The
pendant can also be equipped with an emergency stop. See Figure
6.14.

Keys: ELSTAT
        A.P.
        P.P.
        I/O

This is where the setting of the different process parameters takes
place: paint flow, atomization pressure, and electrostatic voltage.

Keys: MODE
        VER
        SAVE
        END
        POINT
        SPEED
        TCH
        RTCH

This is where the different program routines are executed—for instance, changing between different modes (methods of working) or
between different versions of a particular program. It is also where
you tell the system what kind of programming you want to do:
teach or reteach.

Keys: RUN
        REV

----

*Courtesy of Spine Robotics Corp.

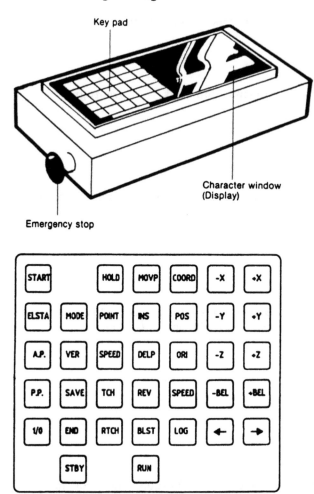

FIGURE 6.14   Programming pendant for the Spine robot.   (Courtesy of Spine Robotics Corp.)

These are keys for moving the robot over a path it has been taught.

Keys: STBY
        BLST

These are keys for axial transposition back to a preprogrammed
starting point.

Keys: MOVP
        INS
        DELP

These keys are used to move, insert, and delete points in a path
that is already taught.

Keys: COORD
        POS
        ORI
        SPEED

When you want to switch coordinate system, speed of movement, or
robot part being programmed, you will use these keys.

Keys: −X
        +X
        −Y
        +Y
        −Z
        +Z
        −BEL
        +BEL

These are the position command keys.   Through them you tell the
system in which direction you want to move the robot.

V;02   5:08   88

E002

The character window provides the operator with information on
where in the program the robot is located, which mode is activated,
the position of the conveyor, and so forth.   It also dispalys error
messages about incorrect commands.

Keys: ▶
        ◀

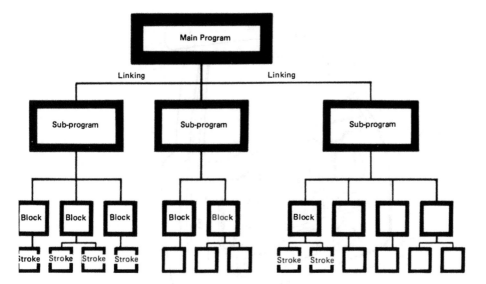

FIGURE 6.15   Programming structure in the Spine robot. (Courtesy of Spine Robotics Corp.)

These keys are used for forward/reverse movement of characters
on the display.

An important part of the work with an industrial robot, and
particularly a robot designed for mechanized production painting,
is the actual learning process.   Teaching should be quick and reliable, and considerable demands are made on exactitude.   The robot
should have the facility for compensating for variations in speed of
the conveyor and the object being painted.   Changing the finished
program should also be a simple operation.   The Spine Spray System meets these requirements.

The Spine programming concept is based on teaching small sequences which together form larger units (subprograms).   The subprograms are linked together to main programs.   The small sequences
are called *blocks*.   The blocks in turn are divided into a number of
working operations which are called *strokes*.   Let us take a car
chassis as an example.   The program comprises spray painting of
the outside of the chassis.   A subprogram covers the painting of
the engine bay, a door, or other component.   See Figure 6.15.

Teaching the robot requires no knowledge of computer programming.   The master program—that is, the program which controls
the control system—is entirely menu-based, and clear instructions

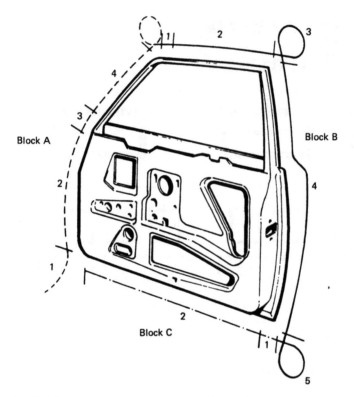

FIGURE 6.16   Program blocks for painting the edges of an automobile door.   (Courtesy of Spine Robotics Corp.)

are presented continuously on the screen and on the pendant display.   The continuous programming work is carried out from the pendant, where the operator is provided with information about the conveyor's position, which blocks or strokes the robot finds itself in, which processor parameters are programmed in, and so forth.

An example of block division when painting the edges of a door is shown in Figure 6.16.   The blocks are marked A, B, and C. The process parameters are steered in the different strokes.   At this point it is decided when and how the paint is to be applied to the door.   The strokes are marked 1, 2, 3, etc. for each of the blocks.   A block can be taught in different *versions*.   These can be tested and compared with each other.   The best version is then saved in the processor and comprises a block.

The operator moves the robot arm in the required path with the aid of the pendant.   By using the *tracking function*, programming can be carried out when an object to be painted is at rest.   Tracking means that the control system is continuously adapted to the robot's movements when the program is played back to changes in speed of the conveyor and the object being painted (position-related tracking). In other words, the operators do not have to make allowance for the conveyor's movements and can concentrate their efforts on suitable robot tracks in peace (Figure 6.17).

The pendant operates the robot's servo system via signals. Through the pendant the operator chooses whether to position the robot or to change the wrist orientation. The operator uses one or more of the three different coordinate systems to move the different parts of the robot.

Editing means making a change in a finished program. Spine's advanced control system provides almost unlimited potential for subsequent correction of movement patterns, speeds, or process parameters. Editing need not apply to the entire program but only to the blocks which are concerned. The new block version can be longer or shorter than the old one. The process parameters can also be adjusted individually for each stroke.

One particular feature is the point editing facility. In an already taught path it is possible to insert new points, move points, and delete existing points. Another feature is the possibility of optimizing the speed in different parts of the program. See Figures 6.18 and 6.19 for further examples of programming.

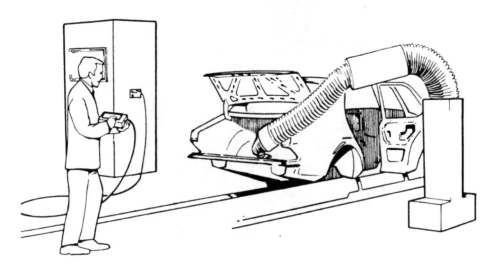

FIGURE 6.17   A robot can be programmed at rest to track a moving conveyor by means of a tracking function. (Courtesy of Spine Robotics Corp.)

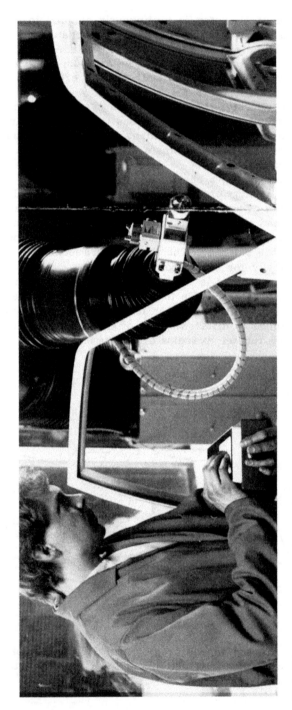

FIGURE 6.18 Programming the Spine painting robot with a teach pendant. (Courtesy of Spine Robotics Corp.)

FIGURE 6.19 Preprogramming phase is extremely important for structuring program sections, blocks, and process information into a complete painting program. (Courtesy of Spine Robotics Corp.)

REVIEW QUESTIONS

The questions below may be used for miniprojects, literature
(library) search activities, and term papers.

1. Discuss the advantages and disadvantages of programming a
    robot by (1) lead-through teaching and (2) walk-through teach-
    ing. Describe two robots which have each of these programming
    methods.
2. Describe the methods of robot motion control. Compare these
    methods to reflect which one provides the best control.
3. Study the enhanced or advanced programming features of two
    robots and indicate the efficiency of these features in
    applications.
4. What is a programming teach box? When is the use of a teach
    box advantageous? Describe the teach box features of some
    assigned robots.

7
Textual Programming

7.1 CLASSES OF TEXTUAL PROGRAMMING

Textual programming languages can be divided into three classes:

1. Joint model
2. Manipulator model
3. Task model

In joint model programming the program is focused at the joint level and each move contains considerable detail about individual actuators. This level is exemplified by lead-through and walk-through robots which use joint model languages and nontextual manual progamming.

In joint model programming the manipulator motion is specified by a sequence of joint motor movements. Typically, an instruction takes the following form:

$$\text{JMOVE } q_1, q_2, q_3, q_4, q_5, q_6$$

The q_i's (i = 1, . . . , 6) represent the joint variables (angles for revolute joints and displacements for prismatic joints). An example of joint level language is ARMBASIC for the Microbot Minimover 5.

The next level comprises manipulator model languages, which are commonly found in most commercial robots. At this level, the program guides the end effector or manipulator tool tip from point to point through Cartesian coordinate space. In many systems the

TABLE 7.1   Illustrative Program for a Pick-and-Place
Task Using a Manipulator-Level Language

---

1. Specification of speed and acceleration

    SPEED: V

2. Definition of gripper locations

    L1: x1, y1, z1, θ1, ϕ1, ψ1

    L2: x2, y2, z2, θ2, ϕ2, ψ2

    L3: x3, y3, z3, θ3, ϕ3, ψ3

    L4: x4, y4, z4

    L5: x5, y5, z5

    L6: x6, y6, z6

3. Definition of gripper movements

    MOVE   L1

    MOVE   L2

    GRASP (FORCE: FY = W)

    MOVE L3   Lift peg above obstacle

    MOVE L4   Approach the hole

    MOVE L5   Insert peg into hole along a straight line

    RELEASE   Release peg

    MOVE L6   Move the gripper away

---

program can be constructed in Cartesian, cylindrical, or tool tip coordinates. Complex action descriptions including time delays, control line toggling, and gripper opening and closing can be incorporated. The conversion from end-effector location to joint variables is automatically done in the robot software. The motion statements programmed by the user are of the following type:

$$\text{MOVE } p_x, p_y, p_z, \theta, \phi, \psi$$

Here p_x, p_y, p_z represent the position of the end effector and θ, ϕ, ψ represent the angles of orientation with respect to the world coordinate frame. Well-structured robot software can also accept

position and orientation with respect to a local coordinate frame.
The robot software will then compute the corresponding global transformations from the specified local transformations.

Examples of end-effector modeling languages include commercial
systems such as VAL for the Unimation PUMA, SIGLA for the Olivetti SIGMA system, HELP for the Digital Electronic Automation PRAG-
MA, the programming systems fo Cincinnatti Milacron and ASEA robots, INDA for Philips robots, and TEACH for the Bendix PACS
system. Other examples found in research facilities are PAL from
Purdue University, ROBEX from the Technical University of Aachen,
and LM from the University of Grenoble. Some systems in this level
were developed earlier and subsequently superseded by more powerful systems. These languages facilitate writing general programs

TABLE 7.2   Robot-Level Languages

Language	Institution
ALFA	General Telephone and Electronics
AML	IBM
EMILY	IBM
HELP	General Electric Corporation
INDA	Phillips
MAL	Milan Polytechnic, Italy
LAMA-S	IRIA, France
LM	IMAG, France
MAPLE	IBM
MCL	McDonnell-Douglas
ML	IBM
RAIL	Automatix
ROBOTLAN	Kawasaki Heavy Industries
RPL	SRI
SIGLA	Olivetti Corporation, Italy
TEACH	Bendix
VAL	Unimation

which can easily be edited to alter, extend, or adapt them when necessary.

Languages such as WAVE, MINI, AL, VAL, and AML generally exemplify traditional procedural languages in syntax and semantics. Other languages such as TEACH, PAL, and MCL show significant departures from the traditional computer programming languages.

Table 7.1 shows an illustrative program for a pick and place task using a manipulator language. Some selected robot-level languages are given in Table 7.2.

7.2   PROGRAMMING THE IRI M50 ROBOT[*]

The IRI M50 robot is programmed on-line via the hand-held move pendant, the system terminal, and the Robot Command Language (RCL), using a combination of "point" and "command" tables which make up RCL. The application program consisting of the point and command table, is maintained in the M 68000 robot control computer, which issues the appropriate commands to the five independent axis control computers.

The point table is the list of points in the robot's coordinates to which an application move command can manipulate the robot. The user teaches the robot each application point by manually moving the robot via the hand-held move pendant to the application point and then pushing the record button. This procedure will cause the existing robot position to be recorded in robot coordinates in the point table. After the point table has been created, the user can create a command table to complete the on-line programming. By entering commands sequentially in the order of execution, which constitutes a command table, the user can program an application. The command table is a list of commands which are executed by the robot controller; it is designed to include the robot commands and basic logic operations, thereby allowing the user complete flexibility in designing the program.

After recording the points and command table for a particular application, the user may move them onto a permanent storage medium via the load terminal or debug the application program via the interactive RCL debug mode and the system terminal. The debug mode allows the user to single-step through the application program. Once the application program has been debugged, it may be saved on the storage medium at the user's convenience. Before running an application program, the user must position the robot to the

---

[*]Reprinted by permission from International Robomation/ Intelligence.

TABLE 7.3   Summary of Robot Command Language

───

Move commands

   Input

CALIB (point #)	Set calibration point (#)
HOMER	Home robot
MOVER	Move robot
SMOVER (point #)	Slow move robot

Gripper Commands

OPEN	Open gripper
CLOSE	Close gripper
WEIGHT (pounds)	Specify weight

Input/output commands

IFIN (#N) (LABEL)	If input #N closed go to (label)
−IFIN (#N) (LABEL)	If input #N open go to (label)
OUT (#N)	Close output
−OUT (#N)	Open output
WAIT (#N)	Wait for input #N to close
−WAITI (#N)	Wait for input #N to open

Set value commands

FINE	Fine positioning
COARSE	Coarse positioning
MSPEED	Modify speed

Logic commands

GO TO (LABEL)	Go to (label)
FLAG (#N)	Set flag (#N)
−FLAG (#N)	Reset flag (#N)
IFFIG (#N) (LABEL)	If flag (#N) set go to (label)
−IFFIG (#N) (LABEL)	If flag (#N) reset go to (label)

Register commands

DELAY (N)	Delay N milliseconds
LCNTR (reg #), (CONSTANT)	Load counter

DCNTR (reg #),
    (LABEL)
	Decrement counter, go to (label)
	    if zero

───

Source: Courtesy of International Robomation/Intelligence.

application calibrate position and then initialize the application program.

A summary of the RCL is given in Table 7.3, which includes robot move commands, logic commands, gripper commands, set value commands, input/output commands, and register commands.

7.3   SURVEY OF SOME ROBOT LANGUAGES

AML is used to control the RS/1 assembly robot, which is a Cartesian arm with linear hydraulic motors and active force feedback from the end effector. A subset of AML is used on the Model 7535 controller by the IBM personal computer.

The TEACH language was developed as part of the PACS system at Bendix Corporation. Two main issues tackled in the PACS system are parallel execution of multiple tasks with multiple devices and definition of robot-independent programs. These two issues have received little attention in other robot programming systems. TEACH has some unique innovations, such as composition programs in partially ordered sequences of statements and specification of all motions relative to local coordinate frames. These features of TEACH are especially important for systems with multiple robots and sensors.

HELP is a commercial language developed at General Electric for their robot products, including the Model A12 Allegro robot system designed for control of multiple Cartesian arms in assembly tasks. The language is Pascal-like and supports concurrent processes to control the two arms in the Allegro system. Motions are defined in terms of the actual point motions, which also specify the XYZ coordinate system. A DEC LSI-11/12 serves as the system controller and an Intel 8080A controls the points.

RAIL was developed at Automatrix, Inc. as a high-level language for the control of both vision and manipulation. It supports interfaces to binary vision and arc welding systems.

MHI (Mechanical Hand Interpreter), the first robot-level programming language, was developed for an early computer-controlled robot, the MH-1, at the Massachusetts Institute of Technology (MIT). The four language primitives in MHI are move, until, ifgoto, and ifcontinue. These primitives have the following features:

1.   "Move" indicates a direction and speed.
2.   "Until" is used to test a sensor for some specified condition.
3.   "Ifgoto" indicates branch to a program label if some condition is detected.
4.   "Ifcontinue" indicates branch to continue action if some condition holds.

TABLE 7.4 Software and Hardware Summary of Some Robot Languages

Language	Geometric data types				Control modes				Motion types			Sensors			
	Frame	Joint angle	Rotation	Transformation	Position	Conveyor tracking	Compliance	Visual	Coordinated Joint	Cartesian	Spline	TV	Force	Limit switch	
AML	X	X			X			X	X			X	X	X	IBM
JARS	X	X	X	X		X	X	X	X		X	X	X	X	DEC PDP-11/34
MCL	X		X	X	X	X				X	X	X		X	IBM 370 DEC PDP-11
RAIL	X				X			X	X	X	X	X		X	Motorola 68000
VAL	X	X			X			X	X	X	X	X		X	DEC KSI-11/03
WAVE		X					X		X	X					PDP 10

The WAVE system was the earliest system designed as a general-purpose robot programming language. Two of the several important mechanisms in robot programming systems, pioneered by WAVE, are

1. The description of positions by the Cartesian coordinates of the end effector
2. The coordination of joint motions to achieve continuity in velocities and accelerations

MCL is an extension of the APT language for numerically controlled machining to robot control. It provides statements for robot motion specification, vision system operation, image modeling, real-time conditional logic, and compile-time language extensions. It supports real-time conditionals in the form of "WHILE-END" and "WHEN-ELSE-END" and allows the programming of flexible relative locations, parallel operations, and macros. MCL was designed for programming work cells in which a number of devices including a robot are controlled by one or more computers. MCL is currently used for the commercial Cincinnati Milacron T3 robot. A summary of software and hardware features of some robot languages is provided in Table 7.4.

7.4   SAMPLE VAL PROGRAMS

Figure 7.1 shows laboratory setups in which the PUMA 600 robot is used for a number of demonstration robot tasks. An industrial palletization application is shown in Figure 7.2. The VAL language used in programming the PUMA has a BASIC-like structure with many new command words added for robot programming. The sample programs presented in Tables 7.5 to 7.8 show some of the key features of the VAL language in programming a robot.

7.5   OBJECT-LEVEL PROGRAMMING

In object-level programming the programs are constructed to model the working environment in a large, sophisticated computer. Artificial intelligence has, to a great extent, led to the development of these languages, which enable moves to be expressed by the positions and motions of objects being manipulated. These world model languages provide symbolic representations of the manipulator, the workspace, and the objects in the task. Examples in this category include the AL language at Stanford University, AUTOPASS at IBM, LAMA at MIT, and LAMA-S at the IRIA in France.

World modeling systems, found mainly in research centers, are goal-oriented and can do more complex tasks and accept more abstract and less detailed instructions. World modeling makes the robot responsible for knowing specific facts about the object it works

 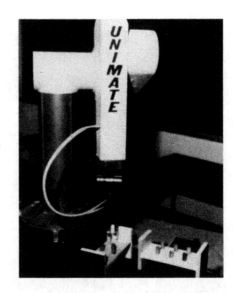

FIGURE 7.1   Laboratory setups for programming a robot for various tasks.

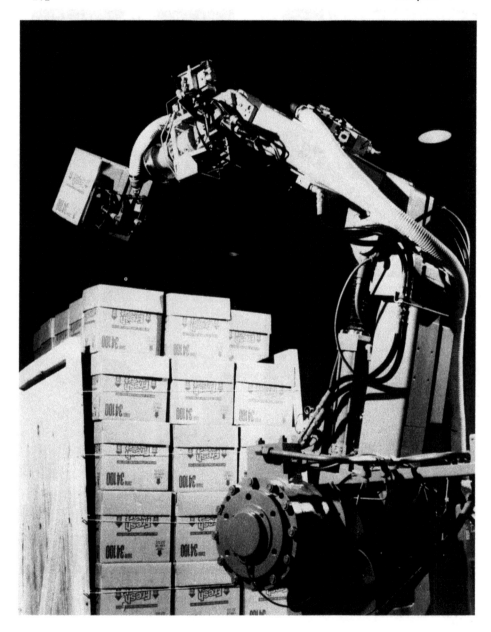

FIGURE 7.2   Palletizing with a robot.   (Courtesy of Cincinnati Milacron.)

TABLE 7.5 VAL Pick-and-Place Program

PROGRAM LAB1

```
 1.  MOVE HOME
 2.  APPRO PICK, 100.00
 3.  MOVES PICK
 4.  CLOSEI 0.00
 5.  DEPARTS 100.00
 6.  APPRO PLACE, 100.00
 7.  MOVES PLACE
 8.  OPENI 0.00
 9.  DEPARTS 100.00
10.  MOVES PLACE
11.  CLOSEI 0.00
12.  DEPARTS 100.00
13.  APPRO PICK, 100.00
14.  MOVES PICK
15.  OPENI 0.00
16.  DEPARTS 100.00
17.  MOVE HOME
END
```

TABLE 7.6 Pick-and-Place Array Program

PROGRAM ROBOT

```
 1.  MOVE HOME
 2.  SET PICK = START
 3.  SET PLACE = END
 4.  SET I I = 1
 5.  10 APPRO PICK, 100.00
 6.  MOVES PICK
 7.  CLOSEI 0.00
 8.  DEPART 150.00
 9.  APPRO PLACE, 150.00
10.  MOVES PLACE
11.  OPENI 0.00
12.  DEPART 100.00
13.  SHIFT PICK BY −45.00, 45,00, 0.00
14.  SHIFT PLACE BY 45.00, 45.00, 0.00
15.  SETI I = I + 1
16.  IF I GT 3 THEN 20
17.  GOTO 10
18.  20 MOVE HOME
```

TABLE 7.7   Palletizing Program

PROGRAM A

  1. SHIFT PI BY −178.50, −51.00, 7.00
  2. SHIFT PL BY −178.50, 5 .00, 0.00
  3. RETURN 0
END

PROGRAM B

  1. SHIFT PI BY −127.50, −51.00, 7.00
  2. SHIFT PL BY −127.50, 51.00, 7.00
  3. RETURN 0
END

PROGRAM PUMA

  1. MOVE HOME
  2. SET PI = PICK
  3. SET PL = PLACE
  4. SETI J = 1
  5. SETI I = 1
  6. 10 APPRO PI, 100.00
  7. MOVES PI
  8. CLOSEI 0.00
  9. DEPARTS 100.00
10. APPRO PL, 100.00
11. MOVES PL
12. OPENI 0.00
13. DEPARTS 100.00
14. SHIFT PI BY 51.00, 0.00, 0.00
15. SHIFT PL BY 51.00, 0.00, 0.00
16. SETI I = I + 1
17. IF I EQ 10 THEN 40
18. IF I EQ 8 THEN 30
19. IF I EQ 5 THEN 20
20. GOTO 10
21. 20 GOSUB A
22. GOTO 10
23. 30 GOSUB B
24. GOTO 10
25. 40 MOVE HOME
26. STOP
27. GOSUB A
END

TABLE 7.8 Alternative Palletizing Program

PROGRAM P

```
 1. MOVE HOME
 2. SET PK = PICK
 3. SETI 0 = 4.
 4. SET PL = PLACE
 5. SETI N = 1
 6. 5 SETI N = 1
 7. 10 APPRO PK, 100.00
 8. MOVES PK
 9. CLOSEI 0.00
10. DEPARTS 100.00
11. APPROX PL, 100.00
12. MOVES PL
13. OPENI 0.00
14. DEPARTS 100.00
15. SHIFT PK BY 50.00, 0.00, 0.00
16. SHIFT PL BY 50.00, 0.00, 0.00
17. SETI M - M + 1
18. IF M GT 0 THEN 20
19. GOTO 10
20. 20 IF N EQ 2 THEN 25
21. IF N EQ 1 THEN 22
22. 22 SHIFT PL BY −175.00, 50.00, 5.00
23. SHIFT PK BY −175.00, −50.00, 5.00
24. GOTO 28
25. 25 SHIFT PK BY −125.00, −50.00, 5.00
26. SHIFT PL BY −125.00, 50.00, 5.00
27. 28 SETI 0 = 0 − 1
28. SETI N = N + 1
29. IF N GT 3 THEN 30
30. GOTO 5
31. 30 MOVE HOME
END
```

---

with. A fairly complete world model is usually required before commands can be accepted. An integrated computer-aided design system may be necessary because of the large amount of information
contained in a world model. However, world model programs are
capable of detecting more errors at the design stages before reaching the factory floor.
    The information needed for a world model of a task includes:

1. Geometric descriptions of all objects
2. Physical descriptions of all objects, such as mass and inertia
3. Kinematic descriptions of all linkages
4. Geometric description of robots
5. Descriptions of robot characteristics such as joint limits, acceler-
    ation bounds, and sensor capabilities
6. Positions of all objects

    The advantage of goal-oriented programming systems is the in-
creased efficiency of the manipulator made possible by off-line programming. For the user, more efficient use is made of robot time
by providing facilities to work at the task level rather than the
manipulator level. Also, the English-like instructions make programming easier. Selected world modelling languages are given in Table
7.9. Sample task-level programs are provided in Table 7.10.
    A brief description of some goal-oriented languages will help to
explain the features of such systems. AL is a general-purpose
language with Algol-like features for describing the motion of the
manipulator and assembly algorithms and the use of sensory information. AL offers interactive programming and debugging facilities for
both assembly and visual processing algorithms, including hot editing with a debugger. Automatic planning of assembly sequences
can be obtained for the user from a description of the final product.
Complete AL programs can be constructed, tested, and debugged
piecemeal.

TABLE 7.9   World-Modeling Languages

Language	Institution
AUTOPASS	IBM
LAMA	MIT
RAPT	Edinburgh University
SIRCH	Nottingham University

TABLE 7.10   Sample Task-Level Programs

Program 1:

   PLACE PEG IN HOLE

Program 2:

   PLACE BRACKET IN FIXTURE S0 (BRACKET
   CONTACTS FIXTURE TOP)

Program 3:

   PLACE INTERLOCK ON BRACKET S0 (INTERLOCK
   BASE CONTACTS BRACKET TOP) AND (INTERLOCK
   HOLE IS ALIGNED WITH BRACKET HOLE)

Program 4:

   SCREW IN NUT ON SHAFT T0 (TORQUE = t0)

Program 5:

   PLACE BEARING SUCH THAT SHAFT FITS BEARING
   HOLE AND BEARING BOTTOM AGAINST SHAFT LIP

AL is the second-generation robot programming language developed at the Stanford Artificial Intelligence Laboratory. It is the result of an attempt to develop a high-level language which provides all the capabilities required for robot programming. The language is based on concurrent Pascal and provides constructs for control of multiple arms in cooperative motion. Both robot-level and task-level specifications are supported by AL. The four major kinds of capabilities provided by AL are

1. Cartesian specification of motions, trajectory planning, and compliance. The capabilities are provided by the WAVE system.
2. Real-time language features such as concurrent execution of processes, synchronization, and on-conditions.
3. Data and control structures similar to an ALGOL-like language, including data types for geometric calculations such as vectors and coordinate frames. The Algol control structures require a user to explain the effects of loops and branches.
4. Support for world modeling exemplified by the AFFIXMENT mechanism for modeling attachments between frames.

Several research groups around the world have used AL, and many commercial arms have been integrated into the AL system. AL is regarded as one of the most complete robot programming systems. It was the first robot language to be a sophisticated computer language as well as a robot control language.

Implementation of AL requires a large mainframe computer, although stand-along portable versions have been announced. Program in AL have been developed and compiled on a PDP-10. The resulting p-code was dowloaded into a PAP-11/45 for execution of run time.

AL most resembles IBM's AUTOPASS, but AL offers interactive programming and debugging facilities for both assembly and visual-processing algorithms, including limited hot editing with the BAIL debugger. Some design features of AL under investigation are speeding up the assembly process, the effect of dimensional tolerance on the probability of correct assembly, and automatic calibration methods for force and torque sensors.

LAMA is a world modeling mechanical assembly language. It allows the user to describe an assembly procedure in typical English statements such as those found in a shop assembly manual. It has extensive text parsing software to translate user commands to robot actions. Additional software components needed for a full LAMA system include a geometric parts design system and a planning system. The computer for LAMA is a PDP-10, and the sensor is a stress-sensing wrist. LAMA is very useful for dealing with all kinds of constrained motion.

AUTOPASS is a powerful world modeling programming language for assembly. It allows commands such as "place object 1 on object 2." It is designed for determining user errors at compile time rather than during execution. Graphic simulation substitutes for interactive debugging in trial runs. The AUTOPASS world model represents assemblies as a graph structure of object part, subcomponent, attachment, and constraint relationships. Basic entities in the graph are points, lines, and surfaces. Statements in AUTOPASS deal with parts, tools, fasteners, and instructions for placement and attachment. Examples of such statements are "operate" for tools, "rivet" for fasteners, and "verify" for a miscellaneous statement. These are translated one at a time in a single pass. The compiler selects optimal grasps for the user and plans hand trajectories to avoid collisions and satisfy the part motion constraints.

7.6   ON-LINE AND OFF-LINE PROGRAMMING

On-line programming proceeds in three steps. First, the coordinates of locations at which some action is to occur are identified.

In point-to-point control these coordinates are generated by the controller and stored whenever the PROGRAM or RECORD button is activated. For continuous path robot location identification is done automatically at a predetermined fixed time interval as the arm is guided along the path. The next step is to state the functions to be performed at a point or along the tool center point path. For some robots these functions are chosen from a manufacturer's list of available functions through special pushbutton selection. Typical point functions are tool manipulation, time delay, and signal. Path functions are velocity specification, search function, and tracking. The final step is definition of the logic of the operation or cycle. This step determines the path of the robot under specified conditions which may be represented by the status of external signals or internal flags.

The on-line teaching procedure is an elaborate and tedious process in aerospace and automotive applications, where hundreds of points must be programmed. For instance, in aerospace industries, many hundreds of holes must be drilled in sheet parts and many rivets are needed for jointing metal parts. In the automotive industry hundreds of spot welds are required in car body assembly. Frequently the coordinates of all the point locations involved in these applications are already available in design drawings existing in a data base. To save time and reduce the redundancy of manually programming such points in an on-line teaching procedure, off-line programming may be used.

In off-line programming the robot is programmed through the use of remotely generated data describing the point coordinates, task function, and cycle logic. Off-line programming eliminates the standard lead-through methods of teaching, and part changes or special jobs can be programmed in advance without disrupting the robot operation. New programs can be loaded and operated without a teach or dry run mode. This means that there is more available pro uctive time for the robot and associated manufacturing line. By using information from a CAD/CAM data base, off-line programming makes it possible for the robot to be more closely integrated with the total manufacturing system. Another feature of off-line programming is that the specification of the robot operation can be combined with the manufacturing process design. Thus decision making and control can be done at a higher level rather than at the shop floor.

In off-line programming, consideration must be given to a number of unique factors that are not present in on-line systems. These factors include procedures for debugging and for interfacing the data base information with the task function and cycle logic. Other problems that must be considered are adjustments for manipulator inaccuracies and alignment of the different coordinate frames.

Adjustment for robot inaccuracies is an important feature of off-line programming. Accuracy is related to the ability of a robot to

go to a commanded point as programmed. This differs from repeatability, which is the expected cyclic variance in the robot position
and the taught point. Repeatability is determined statistically from
test data. As an example, consider a robot which has been programmed to move 40.90 cm. The actual move as measured may be
40.73 cm. On successive runs, the robot arm may move to 40.73
cm. From these data, the accuracy is 0.17 cm and the repeatability
is perfect. This simple illustration shows that the accuracy cannot
be better than the repeatability.

Repeatability can be specified as short-term or long-term repeatability. Long-term repeatability is used to describe robot performance of the same task over many months. Component wear and
aging are the principal factors that influence the long-term repeatability. Short-term repeatability is determined by thermal effects
on electronic and hydraulic components and on the transient condition at start-up.

A second factor which affects the accuracy in off-line programming is the kinematic accuracy, which is related to the tolerances
used in the mechanical construction of the robot. Another factor is
the numeric accuracy, which is related to the word size used in the
computer and the computational algorithms. All these factors contribute to the overall accuracy of an off-line program.

Adjustment of the inaccuracies of a robot arm is done by a calibration technique in the robot workspace. Three-dimensional grid
calibration data can be established typically by using a digital tape
measure, a telescope, and a cross-hair target. The calibration data
can then be used to ensure accuracy of the arm by three-dimensional interpolation techniques.

Alignment of the coordinate reference frames is necessary in
off-line programming because coordinate frames generated off-line
may be skewed to the robot world frame. The zero reference point
for off-line data will not necessarily coincide with the zero reference
point of the robot world frame. In some situations, the part and
robot Cartesian frames may be parallel but offset in some directions.
These alignment errors may be corrected by an alignment frame.
By measuring points on the alignment frame with respect to the
part coordinate frame, vector and matrix values for converting off-
line data to robot coordinates can be determined and stored.

REVIEW QUESTIONS

1. Using a manufacturer's programming manual, describe and compare the programming procedures for a number of current
   industrial robots.
2. Determine the main features and any limitations of the programming language used in each of the robots in question 1.

3. Write a sample program for each of the robots in a typical application involving

    (a) A pick-and-place task.
    (b) Palletizing in two- and three-dimensional arrays.
    (c) Assembly of a shaft in a bearing.
    (d) Continuous motion to generate the uppercase alphabet of the English language.

4. Trace the historical development of robot programming languages, highlighting the ways in which robot applications have influenced the language features.
5. Describe joint motion and Cartesian motion commands in a selected number of industrial robots.
6. Discuss how transformations are used in robot programming languages such as VAL.

FURTHER READING

Ambler, A. P., Popplestone, R. J., and Kempf, K. G., "An Experiment in the Offline Programming of Robots," *12th Int. Symp. Industrial Robots*, Paris, June 1982, pp. 491–502.

Binford, T. O., "The AL Language for Intelligent Robots," IRIA Seminar on Languages and Methods of Programming Industrial Robots, Rocquencourt, France, June 1979, pp. 73–87.

Bonner, S. and Shin, K. G., "A Comparative Study of Robot Languages," *IEEE Comput.*, Dec. 1982, pp. 82–96.

Brinch, P., "The Programming Language Concurrent Pascal," *IEEE Trans. Software Eng.*, Vol. 1, No. 2, June 1975, pp. 199–207.

Darringer, J. A. and Blasegn, M. W., "MAPLE: A High Level Language for Research in Mechanical Assembly," IBM T. J. Watson Research Center, RC 5606, Sept. 1975.

Dijkstra, E. W., "Co-operating Sequential Processes," in *Programming Languages*, F. Genuys, ed., Academic Press, New York, pp. 43–112.

Evans, R. C., Garnett, D. G., and Grossman, D. D., "Software System for a Computer Controlled Manipulator," IBM T. J. Watson Research Center, RC 6210, May 1976.

Falek, D. and Parent, M., "An Evolutive Language for an Intelligent Robot," *Industrial Robot*, Setp. 1980, pp. 186–171.

Feldman, J., et al., "The Stanford Hand-Eye Project," First IJCAI, London, England, Sept. 1971, pp. 350–358.

Finkel, R. A., "Constructing and Debugging Manipulator Programs," Artificial Intelligence Laboratory, Stanford University, AIM 284, Aug. 1976.

Finkel, R., Taylor, R., Bolles, R., Paul, R., and Feldman, J.,
    "AL, a Programming System for Automation," Artificial Intelli-
    gence Laboratory, Stanford University, AIM-177, Nov. 1974.

Franklin, J. W. and Vanderbrug, G. J., "Programming Vision and
    Robotics Systems with RAIL," SME Robots VI, March 1982,
    pp. 392–406.

Geschke, C. C., "A System for Programming and Controlling Sensor-
    Based Manipulators", Coordinated Science Laboratory, University
    of Illinois, Urbana, R-837, Dec. 1978.

Gini, G., Gini, M., Gini, R., and Diusse, D., "Introducing Soft-
    ware Systems in Industrial Robots," *9th Int. Symp. Industrial
    Robots*, Washington, DC, March 1979, pp. 309–321.

Gini, G., Gini, M., and Somalvico, M., "Deterministic and Nonde-
    terministic Programming in Robot Systems," *Cybern. Syst.*,
    Vol. 12, 1981, pp. 345–362.

Grossman, D. D., "Programming a Computer Controlled Manipulator
    by Guiding through the Motions," IBM T. J. Watson Research
    Center, Research Report RC6393, 1977 (declassified 1981).

Kuntze, H. B. and Shill, W., "Methods for Collision Avoidance in
    Computer Controlled Industrial Robots," *12th Int. Symp. Indus-
    trial Robots*, Paris, June 1982, pp. 519–530.

Latombe, J. C., "Equipe Intelligence Artificielle et Robotique, Etat
    d'advancement des recherches," Laboratoire IMAG, Grenoble,
    France, RR 291, Feb. 1982.

Latombe, J. C. and Mazer, E., "LM: A High-Level Language for
    Controlling Assembly Robots," *11th Int. Symp. Industrial Robots*,
    Tokyo, Oct. 1981.

Laugier, C., "A Program for Automatic Grasping of Objects with a
    Robot Arm," *11th Int. Symp. Industrial Robots*," Tokyo, Oct.
    1981.

Lavin, M. A. and Lieberman, L. I., "AML/V: An Industrial Ma-
    chine Vision Programming System," *Int. J. Robotics Res.*,
    Vol. 1, No. 3, 1982.

Lieberman, L. I. and Wesley, M. A., "AUTOPASS" An Automation
    Programming System for Computer Controlled Mechanical Assem-
    bly," *IBM J. Res. Dev.*, Vol. 21, 1977, pp. 321–333.

Lozano-Pérez, T., "Robot Programming," MIT Artificial Intelligence
    Laboratory, AIM Memo No. 698a, April 1983.

Lozano-Pérez, T. and Winston, P. H., "LAMA: A Language for
    Automatic Mechanical Assembly," *5th Int. Joint Conf. Artificial
    Intelligence*, MIT, Cambridge, MA, Aug. 1977, pp. 710–716.

Mujtaba, S. and Goldman, R., "AL User's Manual," Stanford Artifi-
    cial Intelligence Laboratory, AIM 323, Jan. 1979.

Park, W. T., "Minicomputer Software Organization for Control of
    Industrial Robots," *Proc. Joint Automatic Control Conf.*, San
    Francisco, June 1977, pp. 164–171.

Paul, R., "Evaluation of Manipulator Control Programming Languages," *Proc. IEEE Conf. Decision Control*, Fort Lauderdale, FL, Dec. 1979, pp. 252-256.

Paul, R. P., "WAVE: A Model-Based Language for Manipulator Control," *Industrial Robot*, March 1977.

Popplestone, R. J., Ambler, A. P., and Bellos, I., "An Interpreter for a Language for Describing Assemblies," *Artificial Intelligence*, Vol. 14, No. 1, 1980, pp. 79-107.

Popplestone, R. J., Ambler, A. P., and Bellos, I., "RAPT, A Language for Describing Assemblies," *Industrial Robot*, Vol. 5, No. 3, 1978, pp. 131-137.

Ruoff, C. F., "TEACH—A Concurrent Robot Control Language," IEEE COMPSAC, Chicago, Nov. 1979, pp. 443-445.

Salmon, M., "SIGLA: The Olivetti SIGMA Robot Programming Language," *8th Int. Symp. Industrial Robots*, Stuttgart, West Germany, June 1978.

Shimano, B., "The Kinematic Design and Force Control of Computer Controlled Manipulators," Artificial Intelligence Laboratory, Stanford University, Memo 313, March 1978.

Sliver, D., "The Little Robot System," MIT Artificial Intelligence Laboratory, AIM 273, January 1973.

Soroka, B. I., "Debugging Robot Programs with a Simulator," SME CAD-CAM 8, Dearborn, MI, Nov. 1980.

Summers, P. D. and Grossman, D. D., "XPROBE: An Experimental System for Programming Robots by Example," IBM T. J. Watson Research Center, 1982.

Takase, K., Paul, R. P., and Berg, E. J., "A Structured Approach to Robot Programming and Teaching," IEEE COMPSAC, Chicago, Nov. 1979.

Weck, M. and Zuhlke, D., "Fundamentals for the Development of a High-Level Programming Language for Numerically Controlled Industrial Robots," AUTOFACT West, Dearborn, MI, 1981.

8
Robot Vision Systems

8.1 OVERVIEW OF MACHINE VISION*

All pattern recognition systems, including machine vision products,
have three elements: a sensing device, a processor, and an external controller that directs the application (see Table 8.1). Sensors
can be cameras for light, including the X-ray, infrared, and visible
spectra; microphones for acoustic signals; and pressure transducers
for tactile patterns.

The processor element performs its function either off-line or in
real time. Off-line processors address tasks that do not control
operating industrial processes or other timely activities. Since procedures like the analysis of Landsat image data are not time-critical,
traditional pattern processing systems that use general-purpose computers work very well. In fact, most of the software technique for
pattern recognition originated in off-line environments.

Real-time systems, on the other hand, are commonly used in
automation applications and must provide pattern recognition at
speeds high enough to control manufacturing operations. In real-
time situations, general-purpose computers are often taxed by complicated applications because of the complexity of the programs required to perform the recognition process. New technology is needed with performance speeds approaching real-time demands.

Applications for real-time machine vision pattern recognition systems range from straightforward part presence to complicated robot

---

*Adapted by courtesy of Pattern Processing Technologies, Inc.

TABLE 8.1   Components of Pattern Recognition
Systems

Sensors	Processors	Applications
Cameras	Off-line	Inspection
Microphones	or	Recognition
Tactile sensors	Real-time	Measurements
		Guidance

guidance.   Generally, applications fall in several major categories:

Inspection or quality control
Automated part recognition identification
Gauging or dimensional analysis
Visual guidance and control

    Specific tasks that machine vision systems perform in industrial
environments include registration, tracking, measuring, visual quality control, recognition, and parts location.
    True computer-integrated manufacturing will become a reality
only when machine vision systems meet the challenges of real-time
operation and flexibility.   They must be able to interface with a
variety of manufacturing machinery for automated assembly procedures and must be able to react in realistic time frames, since machines do not often follow smooth planned motions or work with
pieces precisely positioned for assembly.
    A machine vision system that can operate in such an environment
ultimately will have to accommodate color, contour analysis, three-
dimensional analysis, and nonstructured environments.   Other im-
provements expected in future systems include large array cameras,
more precise optics, and systems more tolerant of nonuniform
lighting.
    All areas of application of machine vision will expand in the next
few years, but most dramatic will be the growth in areas requiring
precise and fast recognition responses.   Hence, the need exists for
systems not encumbered by complicated programming tasks and over-
taxed general-purpose computers.

Theory of Machine Vision

Machine vision is a type of pattern recognition.   It analyzes input
stimuli (video signals) into classifications or categories—that is, it

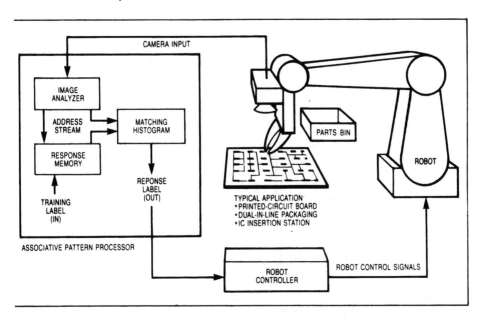

FIGURE 8.1   Pattern recognition's most common commercial use; in machine vision products.   A control guidance application is shown. (Courtesy of Pattern Processing Technologies, Inc.)

matches a current image to a trained image—for the purpose of serving some application or result (e.g., to guide a robot or inspect parts) (Figure 8.1).

A machine vision system includes a camera/translator as a sensing device, a real-time image data processor, and some sort of external controller suitable to the application (Figure 8.2)

Since the camera/translator senses data in a particular scene and converts it to an electronic form acceptable to the data processing component, the quality of the image is of utmost importance. A major concern in this regard is the operational environment. Factors such as light levels, ambient light sources, and particulate matter like oil spray, metal flakes, and dust can change the quality of the image data. A serious problem with many machine vision systems is that they cannot tolerate these conditions.

The camera or image sensor converts the scene into electronic signals. Vision system cameras run the gamut from older black-and-white vidicon-type units to second-generation solid-state units such as charge-coupled devices or charge-injected devices. The second-generation solid-state cameras digitize a scene onto an array of photosensitive cells. The arrays form a pixel grid containing the

SENSING		PROCESSING		APPLYING
IMAGE FORMATION	IMAGE TRANSLATION	DATA ANALYSIS	DATA PROCESSING	TASK CONTROL
• IMAGING • SENSING • LIGHTING	• DIGITAL CONVERSION • MECHANICAL WINDOWING	• BINARY CONVERSION • SEGMENTATION • CONVOLUTION • CONNECTIVITY ANALYSIS • STATISTICAL ENCODING	• REGISTRATION • ELECTRONIC WINDOWING • ORIENTATION • IDENTIFICATION • EXTERNAL REPORTING	• INSPECTION • RECOGNITION • MEASUREMENT • GUIDANCE

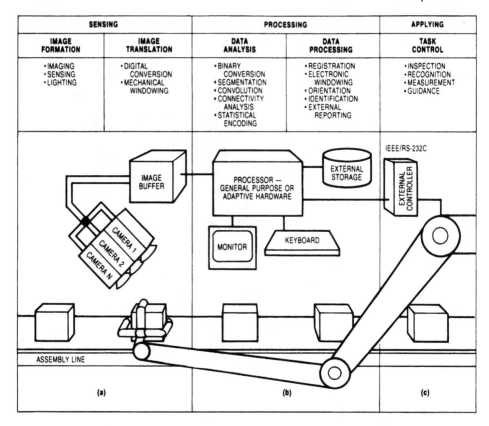

FIGURE 8.2   The three parts of the machine vision process. For
an assembly-line robot, for example, there is a sensor (a) consisting
of a camera and translator, an image-data processor (b) to perform
both analysis and processing, and an external controller (c) for application development.   (Courtesy of Pattern Processing Technologies,
Inc.')

data currently appearing on camera. The solid-state camera comes
in a variety of pixel densities, including 128 × 128 and 256 × 256.
A 512 × 512-pixel camera is also expected on the market in the near
future.
    Once the scene is converted into electronic signals, the second
half of the sensing task occurs—translation of the signals into data
suitable for analysis. This process creates a tremendous amount of
information, which then causes a bottleneck in the image data
processing step. For instance, one image from a 256 × 256-pixel

black-and-white solid-state camera consists of 65,536 pixels. These
can be refreshed 30 times per second. And, if the pixels are interpreted as gray-level values, each one can have up to 256 values.

The image data processor must perform the task of recognizing
the pattern or scene based on the large amount of data provided by
the camera/translator. This recognition process occurs in two
phases: analysis of the image and processing of the analyzed data.

Image analysis is the representation of the translated camera
data by some mathematical model or formula. Since representing all
the combinations possible in every pixel of every scene is not realistic, practical, or always necessary, shortcuts are often taken to
reduce the data volume. One data reduction technique used by a
large number of systems is the conversion of each pixel point into a
binary value. As the name implies, binary systems evaluate each
pixel as black or white. A threshold-adjusting capability often allows users to select the intensity of signal that is to be the black-
white border.

A binary machine vision system reduces the pixel data, but it
also reduces the accuracy of the analysis. Gray-level systems, on
the other hand, can interpret each pixel's value as a specific gray
tone. These systems vary in precision, with each pixel point being
evaluated as 16, 64, or even 256 different values.

Another method of reducing the total data required s to window
the scene—in other words, analyze only a portion of the pixel data
in determining its formula. Since analysis of an entire scene is
often unnecessary for proper recognition, the use of a window can
do a great deal to eliminate unneeded image data.

Segmentation Procedures

The most common means of data reduction is segmentation. This
technique divides the image data into areas of interest and then in-
terpolates surrounding pixels into those areas. There are several
ways to segment an image, but they all make the analysis less accurate. Segmentation procedures are all accomodations to the basic
problem—too much data to be processed and too little computational
power.

Algorithms. Algorithms are a form of segmentation that work with
    an already data-reduced binary image. Algorithms, in general,
    are sets of mathematical models that can be used to describe an
    image. A very popular set of algorithms, the SRI algorithms
    developed at Stanford Research Institute, consist of about 50
    different features that are extracted from a binary image, such
    as the size of blank areas (holes) or objects (blobs), their
    centroids, or their perimeters. Most machine vision systems on
    the market today use SRI algorithms working with binary images.

Neighborhood processing. By treating each pixel value as if it were
part of a group or neighborhood, this gray-level data reduction
technique can change the otherwise complicated image into areas
with well-defined lines. Each pixel's point value is calculated
by considering the value of neighboring pixels. The process
effectively averages the scene into regions or areas.

Convolution. If an image is represented by the rate of light change
per pixel instead of the light intensity, the image will look like
a line drawing because the greatest intensity of a pixel change
is at the boundaries.

String encoding. In run length, string, or connectivity encoding,
the values of the first and last pixel positions of each scan line
are compared to see if they are equal and therefore belong to
the same region. The tables generated by this process also consider the vertical changes in pixel state and, as could be expected, are rather short, because most simple binary images
contain relatively few transition borders.

Once the image data have been analyzed into a mathematical
equivalence, they must be recognized and matched to some trained
or stored value. This step requires a great deal of computational
power for real-time operation, but how difficult it is depends partly
on the reduction techniques used in analyzing the image. If a machine vision system uses a general-purpose computer for this step,
performance levels can be too slow for real-time operation unless the
application involves only very basic procedures, regardless of the
data reduction technique used. Some machine vision systems on the
market also employ parallel processing and pipeline processing to increase computing power. These systems show promise, but the solution may be only temporary. It seems that every time a system
gains a step in processing power, application complexity expands to
fill that capacity.

The final element in every machine vision system is its external
controller, which can interface to robots for guidance control or
connect to inspection-picking systems to find defective parts; the
actual function depends on the application.

Most machine vision systems functioning as components in factory
automation equipment include standard interface controllers that allow
the design engineer to add the machine vision component to other
systems easily.

Problems and Opportunities

There are four basic problems with machine vision systems that use
traditional computer and data processing techniques:

Accuracy. Since the image-sensing element in a machine vision sys-
    tem generates more data than a general-purpose computer can
    handle in real time, many systems employ data reduction tech-
    niques that end up reducing the accuracy of the analysis.
Low speed. The execution of software programs for recognition is
    time consuming. In addition to the base program, there may
    also be several others that must be retrieved from disk storage
    and run for a single event. These can include registration
    routines, programs to determine whether the part is present and
    correctly oriented, and interfacing routines.
Capacity. The industry's initial belief that the advancing computer
    sciences would solve the capacity, speed, and accuracy needs of
    major processing tasks like machine vision has not been justified,
    because image analysis and program complexity are keeping up
    with the increasing capacity and capability of the computer.
Costs. The true cost of a machine vision system based on a gen-
    eral-purpose computer is not confined to the system itself. It
    also includes extensive training and specific applicat on program-
    ming that must be done with each application.

A major jump in processing technology will be necessary if the
machine vision industry is to meet its future expectations. But
faster and better general-purpose computers are not the answer.
Instead, new approaches to data processing are necessary, and
traditional architectures will have to be replaced by a new family of
machines designed specifically for machine vision purposes. In fact,
the new generation of adaptive hardware systems will not be computers at all, but finite-state machines created to perform pattern
recognition.
    One such system, described here, is the associative pattern
processing (APP) technique from Pattern Processing Technologies,
which matches current images to previously trained ones.

Associative Pattern Processing

The basic theory behind the APP technique is simple. A proprietary
statistical encoding system compares the contents of two memories to
find a match or recognition state. The first memory is an image
buffer that contains the pixel data for every point of some current
image, and the second memory contains condensed mathematical equi-
valents of all trained images.
    Condensation of the image data into an actual mathematical equi-
valent or fingerprint is done by proprietary circuitry within the
image analyzer. This circuitry incorporates a statistical phenomenon
of repeating values that occurs when large amounts of data are

samples selectively. The key to the APP's performance level is the technique used to condense the pixel data of an image into a unique "statistical fingerprint."

First, the video signal for a particular image is transferred from a camera to the image buffer. The image buffer is simply a storage device that contains the value of each pixel point appearing on camera. These data are updated every 30 msec, the refresh rate of the camera.

Hardware circuitry within the image analyzer evaluates the data and selects a relatively small set of locations that represent the image in a second memory called a response memory. The locations are generated statistically on the basis of sampling enough image memory points for the image to be recognized. These memory locations are sent to the response memory as a constant address stream; therefore, the statistical fingerprints of the camera images are refreshed in the response memory at the refresh rate of the camera.

During a training session for a particular application, images to be trained are placed in front of the camera, and labels is written into their fingerprint locations in the response memory. When any of these trained images appear during a subsequent working session, the same fingerprints will be generated in the response memory. All you have to do is monitor the memory data lines to identify their labels.

The recognition process takes place at memory access speeds. There is no programming involved in the training process; you simply show an image to the camera, let the fingerprint memory locations drive the response memory, and load the label for each image. In actual operation, the histogram constantly monitors the response memory for trained labels. When recognition occurs. the histogram indicates the label.

The recognition of the trained image is as fast as the response time of the memory. If many images are trained at once with a unique label loaded into each of their unique fingerprints, subsequent appearance of any one of them will produce its label on the read lines of the response memory at access speeds. Thus, recognition speed is independent of the number of images trained into the system.

The associative pattern processing technique is not a computer, but a specialized instrument that can accurately match images on an input camera to trained labels at memory access speeds. The APP and other new architecture systems are and will be designed specifically for the pattern recognition task. The performance levels of these new products are such that they will usher in a new generation of application possibilities only dreamed of with computer-based systems.

For machine vision to become an integral part of computer-integrated manufacturing, it must operate at real-time speeds, be flexible enough to adapt to expanding applications, and be less expensive than current computer-based products. On the other hand, computer-based machine vision systems may not be able to accomodate the complexities of industrial applications. The data volume is too great and the processing requirements too time-consuming. The use of data recognition techniques and volume processing approaches helps the situation but does not solve the problem.

New hardware architecture in finite-state electronic systems is the real solution. Vision instruments using new architecture will give the systems integrator a vision component for artificial intelligence systems. This new generation of machine vision products will be able to perform in real-time industrial environments and still be reasonably priced.

8.2 CAMERAS USED IN MACHINE VISION SYSTEMS*

The various types of television cameras used with machine vision systems generally fall in two main categories: (1) cameras using a light-sensitive vacuum tube for image sensing and (2) cameras using a solid-state sensing device.

Tube-Type Cameras

The most common tube-type TV camera is the vidicon. The vidicon has played a major role in machine visual development and was used in a great many of the earlier machines. With a vidicon or similar device, the image is focused on a photosensitive layer (the target) and a corresponding electrical signal is produced by an electron beam scanning the image target area.

The major features of tube-type cameras are

1. Possible high resolution
2. Possible high sensitivity, depending on the tube selected
3. Image lag due to low response speed
4. Image burn (i.e., permanent or semipermanent retention of image
5. Geometric distrotion and drift (apparent change in position of the image

These effects vary in degree with the type of tube and/or camera used; however, the principles remain the same. Therefore, except in some special cases where sensitivity or resolution dictates otherwise, a solid-state camera should be used.

---

*Adapted by courtesy of Penn Video.

Solid-State Cameras

Charge-coupled devices (CCDs) were invented at Bell Laboratories in the late 1960s. Several companies are currently manufacturing different types of sensors and cameras for industry. Although the techniques are different (hence, the different designations—CCD, CID, MOS, photodiode, etc.), the principle remains the same.

The image sensor elements (photosites) are arranged either in single row as in the case of line scan cameras or in rows and columns as in the matrix (area) camera.

Line Scan Cameras

Typically, a line scan sensor is composed of a row of image sensor elements, two analog transport registers, and an output amplifier. Light energy strikes the photoelements and generates an electrical charge proportional to the light intensity. The charges are transferred to the transport registers by a two-phase clock, amplified, and sent out as a series of amplitude-modulated pulses. These devices come in various degrees of resolution (64, 128, 256, 512, etc.) and are used in optical character recognition and noncontact measurement applications.

The major features of line scan cameras are:

1. They can be selected with high pixel densities (over 3000) and hence are useful in applications such as measurement.
2. Pixel transfer rates can be high in some cases (20 megapixel/sec).
3. Most manufacturers have interface electronics that compress the data and present them to the computer.
4. They produce one-dimensional video information.

The video information from a line scan camera cannot be displayed on a TV monitor. An oscilloscope is normally required for camera setup and information interpretation. This makes it difficult to interface the camera with an untrained operator. In some applications, a scanning mirror is incorporated to extract two-dimensional information. Some considerations in the selection of a line scan camera are (1) resolution, (2) sensitivity, (3) element size and spacing, (4) maximum scan rate, and (5) computer interface. In conclusion, if resolution is not a problem, a matrix (area) camera is easier to set up and operate.

Solid-State Matrix Cameras

The solid-state matrix sensor is composed of multiple rows and columns of image-sensing elements (photosites) and thus can produce two-dimensional images. Several types of cameras are available, the most common being the charge-coupled device, charge injection device (CID), metal-oxide semiconductor (MOS) sensor, and photodiode matrix.

The video output signals from most of these are RS170 compatible, hence they can replace existing vidicon cameras or interface directly with most vision systems. The image signals from the cameras, unlike the line scan type, can be displayed on a TV monitor for ease of setup and interpretation.

The main features of solid-state matrix cameras are:

1. Virtually no geometric distortion.
2. No image drift.
3. Virtually no image lag.
4. No image burn even when imaging highly reflective surfaces.
5. Limited resolution—standard commercial cameras are available with pixel resolutions of 128 × 128, 256 × 256, 320 × 240 and 380 × 488.
6. Blooming characteristics—the overflow of charge to neighboring pixels under high light conditions.
7. "Bad" pixels—some cameras have unworking pixels and pixel columns. Unless a premium is paid for unblemished cameras, the bad pixels are sometimes concealed by pixel averaging techniques.
8. Infrared response—due to the nature of silicon detectors, the spectral response of solid-state cameras is highest in the infrared region (800 to 1100 nm). Since the human eye does not respond to these wavelengths, the camera could produce unpredictable results. The remedy is to use a removal infrared filter to deemphasize the camera's response in that region. Some cameras are so equipped by the manufacturer.

In conclusion, the advantages of solid-state cameras usually outweigh their disadvantages. Therefore, whenever possible they should be utilized in machine vision applications. The main points that need consideration are (1) resolution, (2) sensitivity, and (3) compatibility with the imaging hardware.

8.3 VISION SYSTEM INPUTS[*]

Most vision systems accept several inputs from various sensing devices. These inputs are multiplexed by an electronic switch for subsequent image acquisition and analysis. Some machine vision systems

---

[*]Adapted by courtesy of Penn Video.

are capable of acquiring multiple images simultaneously; however, in most cases they are analyzed sequentially. Processing the image information requires time; therefore, for example, the offer of a 32-camera capability does not necessarily mean a benefit to the user.

Image Acquisition and Processing Time

The timing of TV signals in machine vision systems is based on standard scan rates. Both vidicon and solid-state cameras require 1/60 sec to scan one TV field or 1/30 sec to complete one TV frame (two interlaced fields). In systems with an imaging hardware resolution of up to 320 × 240, only a single field is acquired, taking 1/60 sec. In higher-resolution systems (512 × 480 or higher), one frame is acquired taking 1/30 sec. Assuming single image acquisition hardware and one processor, the processing speed limit would be 30 parts per second for fields and 15 parts per second for frames. (This also assumes that the processing times do not exceed the image acquisition times).

The speeds cited above are available with today's firmware; however, some complex operations require several frame times to process, making the throughput speed considerably lower. Another consideration with high-speed applications is lighting. As stated in the section on lighting, a strobe must be used to "freeze" the motion of moving parts and the user must be sure that the strobe unit is capable of recycling at the required rates.

Resolution

The resolution in machine vision is of two types: spatial resolution and gray scale resolution.

Spatial Resolution. Two factors determine the overall spatial resolution of machine vision systems: (1) the camera resolution and (2) the image digitizer and buffer resolution.

The camera resolution in the solid-state version is the number of elements in the horizontal and vertical directions. In the vidicon type, the camera resolution depends on several factors such as the size of the scanning beam.

In most vision systems the video signals are digitized by an analog-to-digital (A/D) converter and stored in a frame buffer for subsequent computer analysis. The resolution of this device is determined by the sampling rate of the A/D converter and density of the frame buffer. Typically, a digitizer may have a 256 × 256 pixel resolution (i.e., 256 horizontally and 256 vertically). Other examples of pixel resolution are 320 × 240 (yielding a square pixel with standard TV 4 × 3 aspect ratio), 320 × 480, 512 × 480, and so forth.

The spatial resolution determines the smallest detail discernible by the system in a given field of view; for example, in a 320 × 240 system this is 1/320 or 0.3%. However, in most cases the features that are to be detected must be larger than one pixel for reliable analysis (see the section below on resolution accuracy).

Gray Scale Resolution. The gray scale resolution is determined by the possible number of quantized levels assigned by the video A/D converter to each pixel to represent its brightness value.

The numbers usually are 16, 64, and 256 shades of gray. For most applications 64 shades of gray are quite adequate; however, some critical gray scale image analyses may require 256 shades. Therefore the gray scale resolution will determine the smallest change in brightness value of a region detectable by the vision system. As for spatial resolution, relying on one shade change will not produce reliable results (a minimum of 5% to 10% change is required).

Resolution (Spatial) Accuracy

This is an area in machine vision that is often misunderstood by all parties (supplier and user). The confusion usually arises in the measurement applications with vision. The resolution of each subsystem is determined by the components used, for example, the A/D converter on the digitizer hardware. The overall system resolution is determined by the combination of subsystem resolutions (camera, digitizer) and at best is as good as the weakest link in the system. Therefore, the accuracy of a measurement is determined by the overall system resolution and the image quality. Using one pixel to obtain the accuracy required to find an edge might prove unreliable. To obtain reliable accuracy figures, degrade resolution figures several times, depending on the image.

Vision Algorithms

Some of the more common techniques used by vision companies are outlined briefly in the next few paragraphs. Most companies usually focus on one particular technique, which then determines the suitability of their system for certain applications.

SRI and SRI Algorithms

The SRI Vision Module was developed by SRI International and Stanford University and is one of the older techniques used. To process an image the module extracts the edge points (pixel changes from 0 to 1 or 1 to 0). Then the relationships between the edge points are examined for the program to trace and recognize the blobs (regions) in the image. The advantages of such algorithms are: (1) the parts can be recognized regardless of orientation, and (2) the system can be taught by showing it good and bad parts, simplifying the user interface and programming.

The disadvantages are: (1) the algorithms are not suitable for high-speed applications (several parts per second), and (2) high-contrast images are required since the system works with binary (black-and-white) images.

Gray Scale Analysis

For applications involving complex part recognition or surface characteristics (texture, shade, pattern, etc.) gray scale image analysis is required. The gray level is a quantified measurement of image irradiance or brightness. This information is obtained by the digitizer and stored in the frame buffer. The gray scale capability varies with different machine vision systems; however, the figure is generally a power of two: 4, 16, 64, 256. The main advantages of gray scale analysis are: (1) it can be used to examine complex scenes, and (2) with certain algorithms, the system accuracy is less affected by changes in lighting.

Windowing and Pixel Counting

A great many vision applications are solved by dividing the scene into small zones of interest (the windows). The gray scale information is then reduced to binary (black-and-white) information by comparing it to a fixed or calculated threshold. The number of black and/or white pixels is then counted and compared to preprogrammed values for pass/fail decisions. The results for each individual window can be combined logically/mathematically with those for other windows to enable the system to sort, count, and so forth. The window information can be extracted by hardware to speed up the process.

Machine Vision Interface

One of the most important factors in selecting a machine vision system is the degree of computer programming knowledge required by the user for a particular application. If it is established that the user has the capability to program the machine in high-level language, then that route should be taken for the most flexibility.

However, applying machine vision is no simple task, and it requires specialized knowledge and experience. The user should insist on the vision company's support to see the project through. Most companies provide turnkey systems, and they are usually the safest route. Menu-driven systems offer ease of use at some cost in flexibility; however, they enable untrained personnel to utilize machine vision where it counts in the factory.

Machine Interface

This is done through parallel and/or serial (RS232 type) input/output (I/O). The test results, for example, can be communicated to the outside world through individual I/O bits. More complex information such as coordinates for robot guidance or statistical information s communicated through the serial port. For greater noise immunity, the user I/O should be optically isolated.

8.4 APPLICATION OF MACHINE VISION SYSTEMS*

Traditionally, machine vision systems have served as the eyes of robotic systems and the visual element in automated factory inspection systems. Application of machine vision to manufacturing processes can be categorized as follows:

Inspection/quality control
Automated part recognition/identification
Gauging or dimensional analysis
Visual guidance and control

These basic application categories can be discussed in functional terms common to each, such as registration tracking, visual inspection, recognition, location, and dimensioning.

Registration

This task locates a part, a tool, or a robot arm to a known position (Figure 8.3). Because of limited mechanical positioning accuracies or variations in part dimensions, additional correction can be required to position a part accurately. Vision can provide the feedback necessary to correct positioning variations and therefore reduce the precision and cost of the fixed tooling or robot otherwise required for the application. Applications include:

Locating a point for subsequent part insertion or tool application in assembly operation
Locating a point for subsequent visual inspection

---

*Adapted by courtesy of Pattern Processing Technologies, Inc.

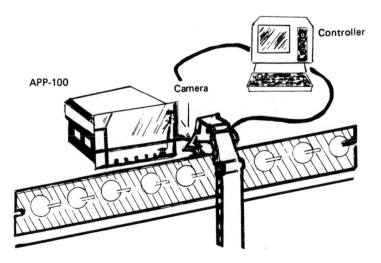

FIGURE 8.3   Robot guidance.   (Courtesy of Pattern Processing
Technologies.)

Locating a pickup point on a part for subsequent pickup by a robot

Tracking.   This task is essentially one of continuous registration.
With vision, the magnitude and direction of part displacement can be
recognized and fed back to the positioning device so that the part
can be followed mechanically.   Operations such as visual inspection
assembly or pickup can then be applied to moving objects.

Dimensioning.   The measurement of the distance between two
points is characteristic of a large percentage of machine vision applications.   This task is accomplished by counting pixels between the
two points.   By use of features such as high-resolution cameras,
windowing, and paging, the APP system can measure distances to an
accuracy of 0.0005 in.   This limitation is due only to the physical
optical constraints of the camera.   Greater accuracies can be achieved
with microscopic lenses.

By using two fixed cameras, APP measurements can be made beyond the field of vision of either camera.   Figure 8.4 illustrates a
situation where the distance Y is determined by measuring W and X,
the distances of the targets from the edge of the camera's field of
view, and then factoring in the known Z, the distance between the
two fixed camera locations.

Visual Inspection.   Visual inspection for quality control or assembly verification is an important industrial task and in many cases
can be automated by machine vision.   Some typical inspection tasks
for which the APP technique is applicable include:

Inspection for completeness or correctness of parts or assemblies
Checking for cosmetic and surface flows
In-process inspection during automated assembly

Visual Recognition. Examples of industrial applications that require visual images to be recognized are described below:

Recognition of the position and orientation of possible pickup points
    on a part so that it may be appropriately accessed by a robot
    arm
Recognition of the presence or nonpresence of a part in machine
    loading and packaging
Recognition of parts in multiple orientations
Recognition of machine malfunctions
Recognition of part types for sorting or packaging

Part Locations. Automatic visual registration of a part is generally applicable only when the part is already relatively close to its
registered position. In many applications, the part may be located
anywhere within a field of view. Thus the first task is to recognize
the approximate position of the part and guide the cameras field of
view toward that position until automatic visual registration can take
over. After visual recognition, other operations can be initiated.

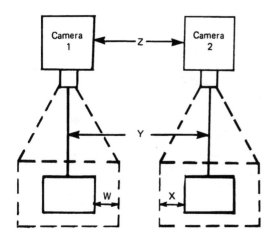

FIGURE 8.4   Dimensioning outside the camera range. (Courtesy of
Pattern Processing Technologies.)

PART SIZING

Application: Inspect brake rotor castings for excessive Total Indicated Runout (TIR). Compare measured data to preset acceptable limits and report accept/reject status for each part.

WIDTH MEASUREMENT

Application: Inspect computer tape reel for disk wobble and width separation. Maintain statistical trend data. Measure and communicate to a PC or host computer to determine accept/reject status.

LABEL INSPECTION

Application: Inspect presence, height position and straightness of labels at 2000 containers per minute. Compare measured data to pre-set acceptable limits and output accept/reject status.

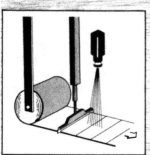

PART TRACKING

Application: Cut panels of uniform width from a continuous roll of raw material. Track leading edge of material. Activate cut-off device when edge reaches predetermined location in sensor field of view.

GAP MEASUREMENT

Application: Maintain constant gap distance between roller drums. Track edges formed by shadows of drum surfaces. Compare measured data to predetermined values and transmit signal to drum adjustment mechanism.

DEPTH MEASUREMENT

Application: Measure depth of holes in castings. Compare measured data to pre-set acceptable tolerances and determine accept/reject status.

FILL HEIGHT CONTROL

Application: Inspect fill height of cans traveling on a conveyor belt at 600 cans per minute. Indicate whether contents are "too high," "okay" or "too low."

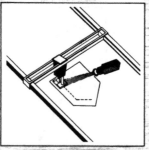

PROCESS CONTROL

Application: Measure edge position of material relative to fixed sewing head. Measured data is sent to x-y controller which adjusts table and facilitates correct part positioning.

INPROCESS INSPECTION

Application: Inspect capacitors at 800 parts per minute. Determine part presence and orientation by comparing the gray scale "signature" of each part to a known standard.

FIGURE 8.5 Application of gauges for inspection. (Courtesy of Honeywell Visitronics.)

242

Industrial Applications

Each of the applications and functions discussed above can be adapted to a variety of industries. The degree to which industries are using machine vision depends on the current performance level of machine vision technology. As systems with improved performance appear on the market, a large number of industries will be using machine vision for a greater number of applications.

The automotive industry has been a leader in incorporating machine vision systems. It is adapting machine vision to virtually all its manufacturing processes including painting, gauging, and inspection. The food industry is finding natural applications for machine vision including clean-room operations and inspection procedures. By use of an X-ray camera, internal foreign matter can be identified and the item rejected.

The medical industry uses machine vision for clean-room procedures and other sterile applications. And if a machine vision product is connected to a microscope, a greater number of inspection, counting, and gauging applications are possible. Package processing can readily adapt machine vision to label reading, distributing, conveying, contents monitoring, and so forth. Gauging, tool inspection, and general automation are just a few uses beyond robotics for which machine vision is used in the machine tool industry.

The chemical industry can apply machine vision systems for fluid level monitoring, clean-room operations, distribution, and hazardous duty areas. The use of machine vision systems in the nuclear industry is a necessity for areas dangerous to workers. The wood products industry can adapt machine vision systems to inspection, label reading, distribution, and guidance for accurate positioning. The electronics industry is another leader in machine vision system applications. Systems are already in use for part location, label reading, part inspection, circuit integrity, and insertion guidance. The steel industry can use machine vision for tasks such as surface inspection, gauging, and measurement. These common industrial procedures illustrate key application areas for machine vision systems. They are common to most industries, and virtually any process that uses them is a candidate for a machine vision product. Machine vision is in its infancy and the next few years should see extraordinary growth into not only the specific industries reviewed here but many others.

Descriptions of other applications are shown in Figure 8.5.

8.5   AUTOMATIC STATIONARY GLASS INSTALLATION SYSTEM

Current trends in manufacturing lend themselves to vision-guided robots for automated work cells. One such system provides the ultimate in flexible automation with its three-dimensional robotic guidance system for stationary glass installation.

The system provides automatic installation of windshields, back-lights, and quarter windows in automobile bodies. Its flexibility permits glass installation into automobiles of all color finishes and all body types, including the new "slippery" cars that require smooth-fitting glass with small tolerances between the glass and body.

With the stationary glass installation system (Figure 8.6) an auto body arrives in a stop station on an unfixtured carrier such as an AGV or Skuk. The location of the car body varies slightly for each installation. Not only is each carrier positioned differently within the cell, but each body has a variable placement on its carrier. In addition, build-to-build variations of the auto bodies can cause positional inaccuracies in body opening locations. With this system, precise positioning of the car body is unnecessary because the vision system locates the window openings regardless of the car's orientation, eliminating the need for expensive fixturing.

After the body arrives in the installation cell, the stationary glass installation system determines the location of the windshield,

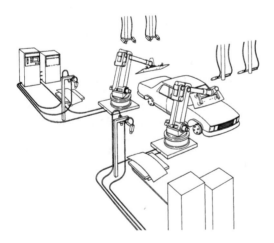

FIGURE 8.6   Automatic stationary glass installation system. (Courtesy of Machine Vision International Corporation.)

backlight, and quarter window openings in a three-space with six degrees of freedom (x, y, z, and rotations about those axes). With this information, the software package for communication with standard robots delivers positional offsets to a robot, which then automatically installs the glass at production line speeds. Because of sophisticated subpixel resolution techniques, the stationary glass installation system is able to achieve installatin accuracies better than 1 mm.

The system capabilities include:

Technology. Industrial gray-scale morphology establishes the location of the low-contrast edges of the body opening on multi-colored, two-tone, and vinyl-topped cars.

Robot interfacing flexibility. The automatic glass installation system works with numerous robots including ASEA, Cincinnati Milacron, GMF, and others.

No part fixturing. The automatic glass installation system locates body openings in unfixtured automobiles in three-space with six degrees of freedom.

Industrial secure camera fixturing. The three-dimensional sensor modules are NEMA 12, enclosed and positioned outside the robot's work envelope. The cameras are placed 6 to 8 ft away while allowing the glass installation system to achieve accuracies better than 1 mm.

Fast and easy calibration. Calibration of the vision system to the robot is performed with a simple command and requires no expensive calibration fixtures. The software supports the intricate communication protocols necessary for secure data transfer.

8.6 FLEXIBLE WORK CELL WITH THREE-DIMENSIONAL ROBOT GUIDANCE

Automation with robots alone provides increased productivity, quality, and flexibility. But installing robots without vision does not allow for assembly of unfixtured components, part-to-part variations, or material handling variations. The following features allow three-dimensional robotic guidance systems to create completely automated work cells:

Gray-scale processing that locates even low-contrast objects accurately in three-dimensional space

Calculation of part location data for six degrees of freedom at assembly cycle rates

Camera and lights located outside the robot's envelope for maximum robot flexibility (typically cameras are 6 to 10 ft away)

No part fixturing beyond existing material handling

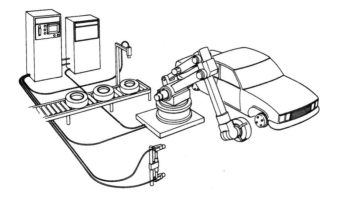

FIGURE 8.7   Application of machine vision in three-dimensional
robot guidance.   (Courtesy of Machine Vision International
Corporation.)

FIGURE 8.8   Quality inspection by a machine vision unit using a
robot for materials handling.   (Courtesy of Machine Vision International Corporation.)

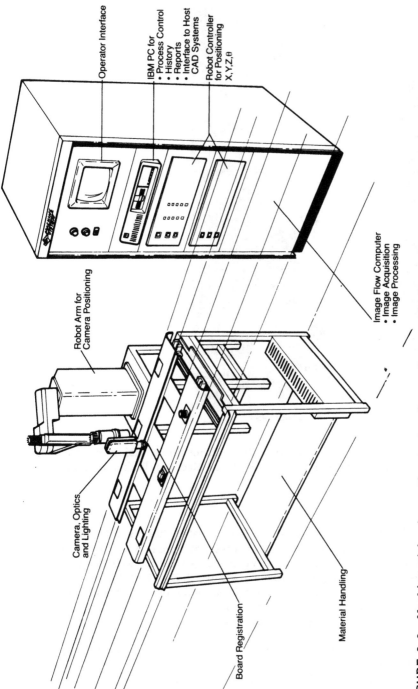

FIGURE 8.9 Machine vision inspection system for surface mounted technology. (Courtesy of Machine Vision International Corporation.)

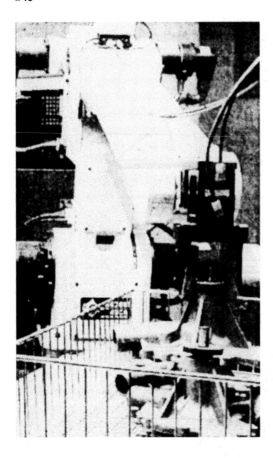

FIGURE 8.10   A robot with three-dimensional vision system picking up an auto fuel pump.   (Courtesy of Machine Vision International Corporation.)

The three-dimensional guidance system begins with binocular vision.   Two cameras, a known distance apart, view a common portion of the workpiece.   Two video images result with a differing relationships of "target features" (Figure 8.7).   Target features are the features in an image whose location resolves the guidance or assembly task the robot is to perform.

Examples of target features include:

The edge of a windshield opening in an automatic windshield installation system

Studs in an automatic wheel mounting system

Frame rivets in an automatic riveting cell

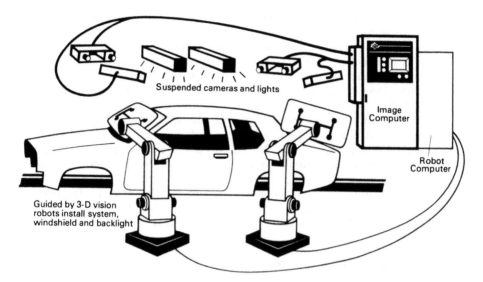

FIGURE 8.11   An MVI vision system for windshield installation.
(Courtesy of Machine Vision International Corporation.)

Parts and rack slots in a robotic rack loader
The fender and quarter panel used for locating an auto body in a
   robotic paint stripping cell

In the process of finding targets, one problem to master is that
of handling variable lighting and variable reflectivity. Parts often
have a wide range of surface textures, from glossy paint to rought
castings. In addition, unless the parts are tightly fixtured, usually
at a high fixturing cost, the objects are often presented to the work
cell with a variable orientation, changing the brightness and contrast in the scene.

Consider the problem of guiding a robot to mount an automobile
wheel on a disc brake hub assembly. The disc rotor is shiny. One
hub assembly may appear to the camera almost pure white, as light
bounces off the rotor straight into the camera. The next hub may
appear substantially darker when the rotor face is rotated differently relative to the camera and light. The important target features
used to perform automatic wheel mounting are the studs. Locating
the studs in this changing scene is virtually impossible for simpler
vision technologies. Morphological gray-scale technology extracts
the studs from the scene and determines their location in the three
dimensions for robotic wheel mounting.

Calibration between a robot and a vision system requires a transformation of vision feature coordinates to robot coordinate data.
This becomes extremely complicated in six-axis robot guidance appli-

cations. The additional flexibility to interface with a wide variety
of robots compounds this transformation problem. The three-dimensional robotic guidance modules perform guidance for the full range
of six-axis robotic mobility: x, y, and z axes, yaw, pitch, and
roll. These modules handle the complex mathematics required to
transform image data into robotic coordinates for a variety of industrial robots.

In addition, the sorfware supports the intricate communications
protocols necessary for secure data transfer. The vision systems
may be controlled by any device in the plant control and communications hierarchy. All interfaces into and out of the system are major
industrial standard, such as MAP.

Other applications of machine vision in robotic systems include
quality inspection (Figures 8.8 and 8.9), materials handling (Figure
8.10), and windshield installation (Figure 8.11).

8.7   PIXIE MACHINE VISION SYSTEM*

PIXIE is a practical gray-level image processing computer designed
for industry. This modular, algorithm-based unit can automatically
identify and locate objects according to their size and shape. It
can also perform quality control and flaw detection functions in real
time.

PIXIE analyzes a scene by using key geometric features to dis-
tinguish specific objects. Cellular automata-based image processing—
accomplished through image algebra—is used to carry out a wide
variety of conceptual operations. For example, the system can identify forms by electronically probing each digital image with selected
geometric shapes (structuring elements) of varying size (Figure
8.12).

When presented with a problem, PIXIE literally asks itself a
series of questions, each designed to describe the desired object in
progressively finer detail. For instance, if PIXIE is looking for
6-32 machine screws that are 1/4 in. long, it might test each of the
objects in its viewing field by applying the following sequence of
questions: Does the object have a silhouette 1/4 in. long? Is it 1/8 in.
thick? Does it have a head? Is the head 1/4 in. wide? Does it have
threads with the correct pitch?

PIXIE will then mark those locations in the viewing field where
the answers to all questions are "yes" and will report their locations.
The standard PIXIE-1000 system includes image processing hardware,
solid-state camera, and a nominal custom algorithm for specific application (Figure 8.13). The modular hardware is based on a new computer architecture developed expressly for image processing. Data
from the video camera are digitized and fed to the processor, where

---

*Reprinted by permission of Applied Intelligent Systems, Inc.

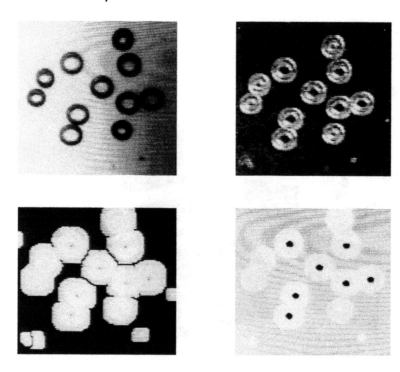

FIGURE 8.12   Real-time image processing.   (Courtesy of Applied
Intelligent Systems, Inc.)

from the video camera are digitized and fed to the processor, where
whole images are subjected to neighborhood transformations.   The
system operates at a nominal speed of 10 MHz.   The system's high
speed, flexibility, and compact size make it ideally suited for modern
industrial applications.   Because of its modular software and hardware design, PIXIE can easily be configured to perform almost any
task that involves visual inspection and identification of objects.
Here are just a few of the many manufacturing areas where PIXIE is
being applied:

Scanning and recording embossed alphanumerics on black rubber (a
        black-on-black problem)
Automatically inspecting liquid vials for various defects
Identifying machine parts and their placement
Checking component accuracy in circuit boards to within 0.001 in.
Identifying areas of grinder burn on heat-treated metal parts
Monitoring seed germination in agricultural laboratories

FIGURE 8.13   The Pixie-1000 machine vision system. (Courtesy of
Applied Intelligent Systems, Inc.)

The standard host software interprets a recognition algorithm by
using the "PIX 4" language. The resulting object code is then sent
to the PIXIE processor cards.

The hardware consists of a number of functionally org nized circuit cards, including:

Special-purpose image processor cards with 10-MHz instruction cycles.
   During each instruction cycle a number of sophisticated opera-
   tions are performed in parallel.
Memory cards cycling at 10 MHz.
Peripheral device interface cards: cameras, monitors, and so forth.
A host controller computer operating in conjunction with widely
   available STD bus components.

PIXIE peripheral interfaces are

GE-2200 128 × 128 video camera

Fairchild CCD-4000 256 × 256 video camera
Television display monitor
DMA interface to STD bus

Standard host controller peripheral interfaces are

CRT terminal, Televideo 925
Keypad, Burr Brown
Disc drive: Commodore 2031
Printer: Centronics, IEEE, or RS-232
User interfaces: RS-232, IEEE (GPIB), 8-bit parallel, or custom

8.8 POYNTING MACHINE VISION COMPONENTS[*]

The Model 208 is a 256 × 256, 8-bit pixel display designed to be
easily interfaced to most parallel or direct memory access (DMA)
computer ports (Figure 8.14). It presents continuous, flicker-free,
RS170 type composite video to display images which can be loaded
into its memory at pixel rates up to 2.2 MSPS. It accepts data on
its 8-bit parallel data lines using only two control signals: an END
OF FRAME (EOF) to indicate the end or start of a new input image,
and a DATA STROBE (DS) to indicate the presence of the next
pixel on the input lines. The polarity of these control lines is strap
selectable for easy interface to most CPU ports.

Model 218 Digital Data Buffer

The Poynting Model 218 is 64K CMOS static memory designed to function as a frame grabber for binary images of up to 64K bytes. Onboard DIP switches allow the user to set the image size from 2 to
64K bytes. The parallel input structure uses only two control signals: EOF to reset for the start of a new block of data, and DS to
indicate the presence of the next byte on the input data lines.
Thus, the input of the Model 218 is easily interfaced to binary image
sources (Figure 8.15).
    Data are transferred from the buffer to the host computer's
parallel or DMA port using only three control lines:

1. GO—from the computer to begin the block transfer
2. BUSY—from the 218 to indicate data fetch in progress
3. TRANS—from the CPU port to indicate acceptance of data

---

[*]Reprinted by permission of Poynting Products, Inc.

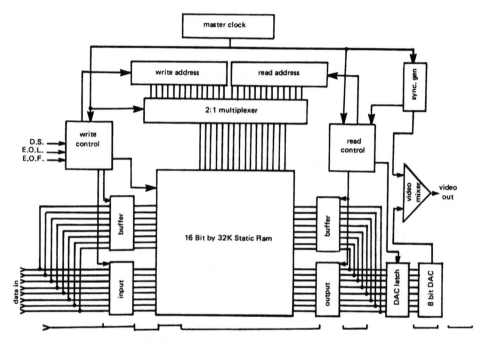

FIGURE 8.14   Functional diagram of Model 208 digital video memory.
(Courtesy of Poynting Products, Inc.)

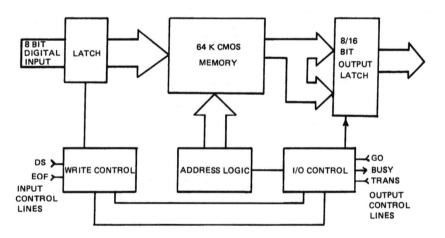

FIGURE 8.15   Functional diagram of Model 218 digital data buffer.
(Courtesy of Poynting Products, Inc.)

The output of the Model 218 is strap selectable for either 8 or 16 bits to make the 218 compatible with either micro- or minicomputers. Data transfer from the Model 218 requires a maximum of 250 nsec per byte in 8-bit output mode and only 375 nsec per byte pair in 16-bit mode. Thus, host CPUs which do not run this fast can access the 218 on a "strobe" basis if desired.

Poynting Binary Image Grabber

Many automated inspection and robotics applications employ binary (1/0) images rather than digital (gray-level) images. The GE TN2505 provides an excellent, moderate-resolution, low-noise, solid-state video image source. The Model 502 is intended to interface the TN2505 to host computers to provide an excellent source of packed binary images.

The Poynting Products Model 502 is a binary image frame grabber and display, designed to accept data from a solid-state imager. The 502 is plug identical to the 2110C, so in most cases the 502 can be implemented by changing the software to process the 384 × 244 pixel image. It accepts data from the solid state and compares the image pixel by pixel to an 8-bit binary value presented by the host

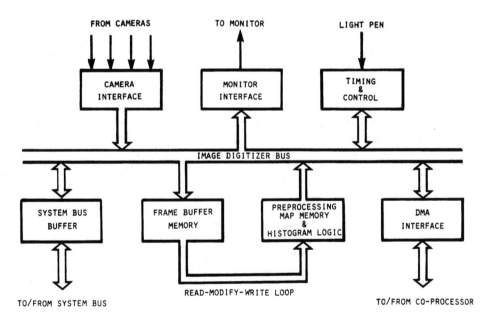

FIGURE 8.16   Poynting Products Model 502 image grabber. (Courtesy of Poynting Products, Inc.)

computer. The image is transformed to a packed binary image on the basis of this comparison and stored in onboard static memory for subsequent transfer to the host. In addition, it has a video output which produces a composite video image of the binary data stored in local memory. It is capable of transferring the image to the host in either a STROBE or HANDSHAKE mode as a strap selectable option. It can also control the INJECT INHIBIT feature of the TN2505 and generate an external strobe signal for flash illumination. Thus, the 502 in combination with the TN2505 presents an effective source of binary images for robotics and inspection applications (Figure 8.16).

8.9 DESCRIPTION OF THE IRI VISION SYSTEM*

The IRI P256 vision system is a high-performance gray-scale image analysis system for industrial application in robotics, inspection, quality control, surveillance, and other areas. The system features a resolution of 256 × 256 picture elements, the picture element being represented by 256 gray levles. Figure 8.17 is a block diagram of the IRI P256.

The P256 has its own dedicated host computer, which is a high-performance microcomputer based on the 16/32 bit microprocessor Motorola MC68000. The wait state free operation of the 12-MHz version of the MC68000 guarantees an execution speed of about 1 million instructions per second (MIPS). The performance of the general-purpose host computer is enhanced by special processors for standard image preprocessing and segmentation functions.

The special processors operate in the *SIMD mode* (SIMD: *S*ingle *I*nstruction–*M*ultiple *D*ata stream) under supervision of the host computer. They are designed to perform a comprehensive multitude of standard image analysis routines at a rate of 2 million to 3.25 million picture elements per second (a 256 × 256 gray-level matrix represents approximately 65,000 picture elements). If the image analysis procedures performed by the special processors were programmed on a conventional computer, the computer would need to have a performance of tens of millions of operations per second to attain the execution speed of the special processors.

The high performance offered by P256 at very low cost is the result of an innovative utilization of very large scale integration (VLSI) technology. Thus, the P256 provides a performance high enough to satisfy real-time requirements in industrial inspection and robotics applications.

---

*Reprinted by permission of International Robomation/Intelligence.

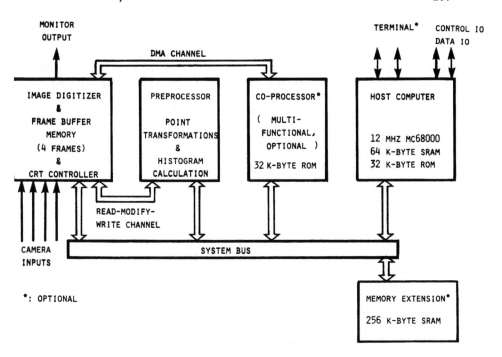

FIGURE 8.17   Block diagram of the P256 digitizer.   (Courtesy of International Robomation/Intelligence.)

The 256, being a gray-scale image analysis system, offers a host of powerful image preprocessing and segmentation routines that cannot be performed by a binary system, that is, a system in which the gray-level information of the picture elements is lost at picture digitalization (i.e., from the beginning). Examples of algorithms performed by the P256 which could not be performed by a binary vision system are

Edge detection by calculation of gradients
Application of the convolution operator, for instance, for noise
    filtering (smoothing), sharpening, edge detection, and curve
    following
Image subtraction and addition
Histogram analysis for optimal threshold setting

The architecture of the P256 is shown in Figure 8.18. In its basic configuration the system consists of an image digitizer and preprocessor, host computer, and coprocessor (optional).

The image digitizer has four camera inputs which can be selected under program control, a frame buffer memory of 256K bytes capable of accomodating four frames (images), and a monitor output. The preprocessor is inserted into the read-modify-write loop of the frame buffer memory. Consequently, its function are performed "on the fly," for instance, during image input. The coprocessor communicates with the frame memory via a special DMA channel that includes the special address generation schemes required for performing a variety of different operations in the SIMD mode. The host computer has its own onboard SCRAM and ROM. A simple program development environment can be provided by adding a terminal, a printer, and a load terminal to the host computer.

Digitizer

The purpose of the P256 digitizer is to digitize gray-scale images supplied by a television camera and store the digitized picture in a *frame buffer memory*. See Table 8.2 for specifications. Furthermore, a monitor output allows the user of the system to view digitized images on a CRT screen and thus monitor the effects of the various steps of an image analysis procedure.

The P256 digitizer has four camera inputs which can be addressed by the host computer. Two things can be done with the

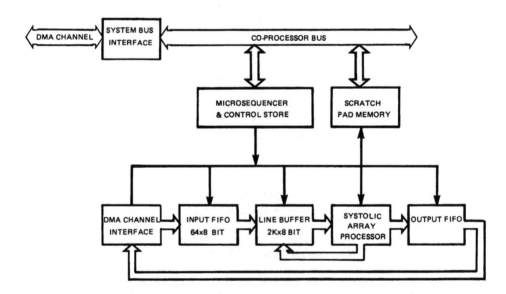

FIGURE 8.18   Block diagram of the P256 coprocessor. (Courtesy of International Robomation/Intelligence.)

TABLE 8.2   Specifications of Digitizer.

Camera	Any vidicon or solid-state camera can be used that operates according to either the EAI line TV standard. The camera must have the RS-170 interface and an input for external synchronization
Image representation	Gray-level matrix with 256 × 256 picture elements and 256 gray levels per picture element
Analog-to-digital converter	8 bits, sampling rate 6 MHz
Digital-to-analog converter	8 bit converter
Frame buffer memory	256K bytes of dynamic memory
Read-modify-write channel	32 bits wide, DMA address generation logic
Co-processor DMA channel	32 bits wide, DMA address generation logic
System buffer interface	Interface to the IRI bus
Light pen input	Yes

Source: Courtesy of International Robomation/Intelligence.

video picture supplied by the selected camera (1) view the live picture on the monitor and (2) digitize the picture and store its gray-level matrix representation in one of the four *pages* of the frame buffer memory.

The 256K byte frame buffer memory is in the address space of the host computer and, therefore, directly accessible to the programs in the host computer simply by referencing.

In addition, the P256 digitizer comprises two special DMA channels: the *read-modify-write channel* for image input and preprocessing, and the *coprocessor DMA channel* for SIMD operation of the coprocessor.

As shown in Figure 8.18, the P256 digitizer consists of the following functional units:

Camera interface with video signal former and analog-to-digitial converter
Monitor interface with digital-to-analog converter and CRT controller

TABLE 8.3   P256 Host Computer Specifications

Central processing unit	Microprocessor Motorola MC68000, 12-MHz version
Performance	Wait state free mode of operation, leads to approximately 1 million instructions per second if amount of multiplications/divisions is small
Memory	64K bytes SRAM, onboard, 70-nsec access cycle[a]
	256K bytes on extension board (optical), 70-nsec access cycle
	32K bytes EPROM, onboard, to accommodate the RT/M real-time monitor and some basic utility routines[a]
	32K bytes EPROM, on coprocessor board, to accommodate routines of the P256 subroutine library
Bus	IRI system bus, asynchronous, 23 address lines, 16 data lines, 14 control lines
I/O interface	RS-232 bit-serial terminal link, RS-422 bi-sync data link, 8+2 bit control output
Interrupt logic	Eight single-line, vectored interrupts with automatic vector generation
Timers	Three independent timers, 16 bits
Self-test logic	Dedicated self test hardware, checking interrupt logic onboard RAM onboard EPROM RS-232 serial I/O

[a] SCRAM, static random access memory; EPROM, erasable programmable read-only memory.
Source: Courtesy of International Robomation/Intelligence.

Frame buffer memory
Read-modify-write channel
Coprocessor DMA channel
System buffer interface
Timing and control logic

Coprocessor

The P256 coprocessor performs a variety of powerful operations for
image preprocessing and image segmentation at high speed. It is
the coprocessor that provides the P256 vision system with the real-
time capabilities required in many applications of robotics and inspection. The P256 coprocessor is optional; any function performed
by the coprocessor can also be performed by a programmed routine
in the host processor—however, one to two orders of magnitude
slower. See Table 8.3 for host computer specifications. Therefore,
whether the coprocessor is needed depends on the application.
     Operations performed by the P256 coprocessor are

Copying of an image and scaling
Image subtraction or addition and scaling of the result
Calculation of the Roberts gradient of an image
Calculation of the first three moments of an image
Convolution of an image with a 3×3, 5×5, or 7×7 coefficient
     matrix

     The P256 coprocessor is a microprogrammed special-purpose
processor consisting of the following units:

Microprogrammed controller
Systolic array processor and buffer for coefficients and intermediate
     values
High-speed scratch pad memory
DMA channel interface, including address generation logic

     The innovative use of the systolic array principle results in a
proprietary processor in which high performance is obtained at low
cost. The microprogrammability of the coprocessor offers the potential of adding new operations or modifying the existing ones.
     All operations of the coprocessor are performed in a pipelined
fashion, the elementary pipeline clock being 50 sec. The resulting
maximum processing speed of the co-processor is 40 million arithmetic operations per second.

8.10   DESCRIPTION OF RANK VIDEOMETRIX
ROBOTIC OPTICAL INSPECTION SYSTEM*

The automatic robotic optical inspection system consists of a high-
performance host computer, a state-of-the-art video inspection system, an advanced precision robot, and the associated hardware and
integration software. The host computer controls the overall operation of the system, performs the statistical analysis of the test results, and is the normal operator interface. Inspection results are
displayed on the host computer CRT and, if desired, printed on the
host computer printer. Lot summaries, statistical data, and reject
data are the normal printouts for permanent records.
     The robotic inspection consists of the following equipment:

1.   Host computer and associated peripherals
2.   Video inspection system with associated cameras, optics, lighting
     sources, camera mounts, television monitor, and other associ-
     ated hardware
3.   Robot with associated controller, teach terminal, and other
     associated hardware
4.   Robot end effector to provide the material handling capabilities
     during the inspection process

     The robotic inspection consists of the following software:

1.   Integration software to integrate the host computer, video sys-
     tem, robot, and associated external hardware
2.   Video inspection firmware to perform the actual inspection
3.   Calibration software for the system
4.   Statistical software for performing inspection lot analysis
5.   Menu-driven host software for the inspection process
6.   Formatted inspection results for the printer which list the re-
     quired dimension and tolerance, actual value, and difference, if
     the dimension is out of tolerance.

Host Computer System

The host computer is the Hewlett Packard Model 216 computer (or
equivalent). The central processing unit of the Series 200 computer
is the Motorola 68,000 microprocessor, which is a high-performance
16-bit microprocessor with a 32-bit internal data structure. The

---

*Adapted by courtesy of Rank Videometrix, Inc.

high performance of this computer is a result of the 16/32 micro-processor operating at a clock frequency of 8 MHz. The host computer has 512K of memory, a 9-in. freestanding CRT for menus, a detached keyboard with special function keys, and a built-in IEEE-488 (HP-IB) parallel interface.

Computer peripherals include

1. Dual 3-1/2-in. flexible disc drive, Model 9121D, with eight blank diskettes furnished for program and data storage.
2. 80-column dot matrix printer (HP Model 82906A) which prints at 160 characters/sec bidirectional. The printer has a 2K buffer and is furnished with one box of paper.
3. A serial port for communications with the robot and the video subsystem.

System integration software for the host computer, video subsystem, robot, and associated peripheral equipment.

Optical Inspection System

The overall video system includes the following items:

1. A Rank Videometrix advanced video processing system.
2. A video multiplexer to facilitate a multicamera system.
3. A television monitor for displaying the current selected camera. The monitor displays one image at a time or one camera in a continuous scan mode.
4. An RS-232C communications port to interface with the host computer.

The optical system design is one of the most important issues in the performance of the optical inspection system. To ensure repeatable inspection results, proper attention must be given to the optical design. Each optical inspection station, in addition to the video system, includes the following items:

1a. A MTI/DAGE precision industrial grade vidicon camera or
1b. A solid-state matrix camera, typically the Sony XC-37 CCD camera. The Sony camera is a high-resolution solid-state camera with an array of 491 × 388 pixels.
2. Appropriately sized lens.
3. Extension tubes, if required.
4. Light source to illuminate the part appropriately in order to obtain a quality image.
5. Camera mounting and mounting for the light source.

Digital Image Processor

The video subsystem is the Rank Videometrix digital image processor Model 100. The Model 100 contains a high-performance 16-bit Intel 8086 microprocessor with the 8087 math coprocessor. The attached digital image processor product description delineates in detail the capabilities of the video subsystem.

Briefly, the key features of the digital image processing Model 100 can be summarized as

1. Compatible with any RS-170 tube or solid-state camera.
2. Up to four cameras with the standard video multiplexer.
3. 256 levels of gray.
4. Input intensity transformation lookup tables for performing binary, inverse binary, semibinary, or inverse semibinary transformations on the incoming video data.
5. A $512 \times 512 \times 8$ frame buffer for storing a gray-level image.
6. Output intensity transformation lookup tables for performing binary, inverse binary, semibinary, or inverse semibinary transformations on the video data to the system television monitor.
7. Comprehensive set of feature extraction commands for the extraction of object edge data in the presence of noise and dust. Statistical analysis is performed on the extracted data to provide the best estimate of the true edge of the object.
8. Intensity normalization of the image to enhance the quality of a poor-contrast image for better inspection results.

Robot

The robot selected for this application is the SEIKO Model RT-3000. The key features of this robot are

1. High repeatability of ± 0.001 in. in the entire workspace
2. Composite speed of 55 in./sec
3. Simultaneous four-axis movement
4. Large workspace, namely
    a. 290° plane rotation
    b. 11.8 in. horizontal stroke
    c. 4.7 in. vertical stroke
    d. 290° grip rotation
5. Teach terminal for direct robot control
6. Input/output for
    a. 16 parallel I/O at the robot controller
    b. four data lines
    c. eight input and eight output at the end effector
    d. external emergency stop
    e. external stop/start

     f.   RS-232C port to the host computer
     g. parallel printer port
7.   DARL—a user-friendly robot language similar to Basic
8.   Storage of 400 absolute points in the overall workspace
9.   Palletizing software for picking and placing parts that are placed in arrays such as pallets
10. Frames for defining up to 10 different coordinate systems
11. Ten levels of nested subroutines
12. Search algorithm for finding parts and defining parts in the workspace
13. Payload of 11 lbs

The robot is integrated into the host computer and the video system for a complete robot inspection system.

A number of other robots on the market can be used as an inspection robot. Another robot in the Seiko product line is the RT2000. The RT2000 uses the same electronics and software but is a smaller, faster, more accurate robot. The key differences are

1.   High repeatability of ± 0.0004 in. in the entire workspace
2.   Composite speed of 63 in./sec
3.   Large workspace, namely
     a. 290° plane rotation
     b. 9.84 in. horizontal stroke
     c. 3.94 in. vertical stroke
     d. 290° grip rotation
4.   Payload of 2.2 lbs

System Repeatability

The repeatability of the system is based on the system design. If the robot actually moves the part being inspected or moves the camera over the part, the largest source of error is the robot itself. The Seiko RT 3000 robot has a repeatability specification of ± 0.001 in. for the entire workspace.

On the other hand, if the robot is just the material handler for the system and the inspections are being performed, say, on a precision XYZ staging, then the repeatability of the system is independent of the robot itself and is a function fo the staging and the associated optical inspection system.

The basic resolution of a vision system is the field of view divided by the number of pixels. The repeatability of the video measurement is typically better than the basic resolution because of the inspection methodology. With a camera, there are over 200,000 data points (pixels) for each measurement. Therefore, using statistical methods, the true edge can be predicted in the presence of noise to subpixel repeatability.

Other considerations include ambient temperature variations, equipment vibration, and so forth.

For a system where the robot presents the part to the camera(s) or moves the camera over the part, the repeatability of system will be typically less than ±0.002 in. for each reading point. The repeatability of system should actually be much better than ±0.002 in. when making small moves between measurements.

Of course, if another robot is used the system repeatability will be affected. If the Seiko RT2000 is used, the repeatability will actually be improved; the repeatability of the RT2000 is ±0.004 in. for the entire workspace.

System Throughput

The throughput depends on the number of optical inspections required and the amount of material handling by the robot. The typical inspection time is 1 sec per image. One image may have a number of inspection points; multiple inspections in the same field of view occur very rapidly. The 1 sec per inspection is for a short move to an inspection point near the last inspection point, which is usually the case.

This time can be reduced by implementing a statistical sampling plan. For example, if it is inherently unlikely for dimension A to be out of tolerance if dimension B is within tolerance, then A need only be inspected occasionally to verify the correlation between A and B. Condition inspection reduces the cycle time. If dimension B is within a specified range of values, dimension A can be skipped. Usually, optimization of the inspection cycle occurs after a history is established on a part of process.

FURTHER READING

Vision Sensors

Agin, G. J., "Vision Systems for Inspection and Manipulator Control," *Proc. 1977 Joint Automatic Control Conf.*, June 1977, pp. 132–136.

Agrawal, A. K. and Epstein, M., "Robot-Eye-in-Hand Using Fiber Optics," *Proc. 3rd Int. Conf. Robot Vision Sensory Controls*, Cambridge, MA, Nov. 1983, pp. 257–262.

Alvertos, N. and Hall, E. L., "Omnidirectional Viewing for Robot Vision," *Proc. 3rd Int. Conf. Robot Vision Sensory Controls*, Cambridge, MA, Nov. 1983, pp. 309–318.

Baird, H. S. and Lurie, M., "Precise Robotic Assembly Using Vision in the Hand," *Proc. 3rd Int. Conf. Robot Vision Sensory Controls*, Cambridge, MA, Nov. 1983, pp. 533–540.

Beni, G., Hackwood, S., and Rin, L., "Dynamic Sensing for Robots—An Analysis and Implementation," *Proc. 34rd Int. Conf. Robot Vision Sensory Controls*, Cambridge, MA, Nov. 1983, pp. 249–256.

Bolle, R., Cernuschi, B., and Cooper, D. B., "Fast Parallel Image Processing for Robot Vision for the Purpose of Extracting Information about Objects," *Proc. ICASSP '83, IEE Int. Conf. Acoustics, Speech Signal Processing*, Boston, April 1983, Vol. 1, pp. 126–127.

Brady, M., "Seeing Machines: Current Industrial Applications," *Mech. Eng.*, Vol. 103, No. 11, 1981, pp. 52–59.

Braeuer, R., "Videomat—A TV Analysing System for Process and Quality Control," *Siemens Power Eng.*, Vol. 4, No. 2, 1982, pp. 76–79.

Buckley, S., "Phase Monitoring: Continuous Wave Acoustic Sensors for Robots," *Proc. 2nd Int. Computer Eng. Conf.*, San Diego, Aug. 1982, Vol. II, pp. 43–48.

Casasent, D. and Cheatham, L., "Hybrid Optical/Digital Moment-Based Robotic Pattern Recognition System," *Proc. SPIE Int. Soc. Opt. Eng. (USA) Robotics and Industrial Inspection*, San Diego, CA, Aug. 1982, Vol. 360, pp. 105–111.

Chiang, M. C., Tio, J. B. K., and Hall, E. L., "Robot Vision Using a Projection Method," *Proc. 3rd Int. Conf. Robot Vision Sensory Controls*, Cambridge, MA, Nov. 1983, pp. 113–120.

Chinlon, L., et al., "High-Power Lasers and Optical Waveguides for Robotic Material-Processing Applications," *Bell Syst. Tech. J.*, Vol. 62, No. 8, 1983, pp. 2479–2492.

Chiu, M. Y., "A High-Performance Vision System for Robotics," *Siemens Forsch. Entwicklungsber.*, Germany, Vol. 13, No. 3, 1984, pp. 147–150.

Corby, N. R., "Machine Vision Algorithms for Vision Guided Robotic Welding," *Proc. 4th Int. Conf. Robot Vision Sensory Controls*, London, Oct. 1984, pp. 137–148.

Dandiker, R., Hess, K., and Sidler, T., "Hybrid Coherent Optical and Electronic Object Recognition," *Appl. Opt.*, Vol. 22, No. 14, pp. 2081–2086.

Dulin, S. K., "Variant of Algorithmic Programming of a System of Visual Perception of an Assembling Robot," *Eng. Cybern.*, Vol. 19, No. 5, pp. 96–107.

Dvorovkin, V. E. and Pankov, V. A., "Device for Data Processing in an Engineering Vision System," Report No. NRC/CNR-TT-2065, 1983.

Erlebach, M. B., "High Speed Cost Effective Vision System for Robots," *Proc. 7th Brit. Robot Assoc. Annu. Conf.*, Cambridge, MA, May 1984, pp. 147–154.

Filo, A., "Concepts of Biostereometrics Applied to Robotics," *Proc. SPIE Int. Soc. Opt. Eng. (USA) Biostereometrics '82*, San Diego, Aug. 1982, Vol. 361, pp. 346–52.

Fukui, I., "TV Image Processing to Determine the Position of a Robot Vehicle," *Pattern Recognition*, Vol. 14, No. 6, 1982, pp. 101–109.

Gelbgras, D. and Bogaert, M., "Grabbing Moving Objects with Visual Feedback," *Sensor Rev.*, Vol. 3, No. 2, 1983, pp. 76–80.

Gregory, P. J. and Taylor, C. J., "Knowledge Based Models for Computer Vision," *Proc. 4th Int. Conf. Robot Vision Sensory Controls*, London, Oct. 1984, pp. 325–330.

Gruver, W. A., et al., "Off-Line Robot Vision System Programming by Computer-Aided Design," *Proc. 8th Int. Conf. Industrial Robots*, Detroit, June 1984, Vol. 2, pp. 14.13–14.26.

Hill, J. J., "Vision Guided Assembly: A Case Study from the High Power Semiconductor Industry," *Proc. 7th Brit. Robot Assoc. Annu. Conf.*, Cambridge, UK, May 1984, pp. 139–146.

Hobrough, G. and Hobrough, T., "Stereopsis for Robots by Iterative Stero Image Matching," *Proc. 3rd Int. Conf. Robot Vision Sensory Controls*, Cambridge, MA, Nov. 1983, pp. 285–294.

Ikeuchi, K., et al., "Picking up an Object from a Pile of Objects," MIT Report No. AI-M-726, 1983.

Ishii, M., et al., "A New 3D Sensor for Teaching Robot Paths and Environments," *Proc. 4th Int. Conf. Robot Vision Sensory Control*, London, Oct. 1984, pp. 155–154.

Kanade, T., "Geometrical Aspects of Interpreting Images as a Three-Dimensional Scene," *Proc. IEEE*, Vol. 71, No. 7, 1983, pp. 789–802.

Kanade, T. and Sommer, T. M., "An Optical Proximity Sensor for Measuring Surface Position and Orientation for Robot Manipulation," *Proc. 3rd Int. Conf. Robot Vision Sensory Controls*, Cambridge, MA, Nov. 1983, pp. 301–308.

Kelley, R. B., "Heuristic Vision Algorithms for Binpicking," *Proc. 14th Int. Symp. Industrial Robots/7th Int. Conf. Industrial Robot Technol.*, Gothenburg, Sweden, Oct. 1984, pp. 599–610.

Kelley, R. B., et al., "Robot System Which Acquires Cylindrical Workpieces from Bins," *IEEE Trans. Syst. Man. Cybern.*, Vol. 12, No. 2, pp. 204–213.

Kent, E. W., "Hierarchical, Model-Driven, Vision System for Sensory-Interactive Robotics," *Proc. Compsac 82 6th Int. Conf.*, Chicago, Nov. 1982, pp. 400–409.

Komarov, V. M., "Calculation of Parameters of the Optical System of the Image Transducer of an Adaptive Industrial Robot," *Izv. Vyssh Uchebn. Priborostr.*, Vol. 26, No. 3, 1983, p. 95 (in Russian).

Kothari, T., "The Role of Vision System in Robotic Applications,"
    8th Int. Conf. Industrial Robots, Detroit, June 1984, Vol. 2,
    pp. 14.63–14.72.
Kuroda, S., Jitsumori, A., and Inari, T., "Ultrasonic Imaging Sys-
    tem for Robot Using Electronic Scanning Method," *Proc. '83 Int.
    Conf. Advanced Robotics*, Tokyo, Sept. 1983, pp. 187–194.
Ledoux, O. and Bognert, M., "Pragmatic Approach to the Bin Pick-
    ing Problem," *Proc. 4th Int. Conf. Robot Vision Sensory Con-
    trols*, London, Oct. 1984, pp. 313–324.
Loughlin, C. and Morris, J., "Applications of Eye-in-Hand Vision,"
    Proc. 7th Brit. Robot Assoc. Annu. Conf., Cambridge, UK,
    May 1984, pp. 155–164.
Loughlin, C., et al., "Line, Edge and Contour Following with Eye-
    in-Hand Vision System," *Proc. 14th Int. Symp. Industrial
    Robots/7th Int. Conf. Industrial Robot Technol.*, Gothenburg,
    Sweden, Oct. 1984, pp. 553–560.
Luo, R. C., Suresh, S., and Grande, D., "Sensors for Cleaning
    Castings with Robot and Plasma-Arc," *Proc. 3rd Int. Conf.
    Robot Vision Sensory Controls*, Cambridge, MA, Nov. 1983,
    pp. 271–278.
McLauchlan, J. M., et al., "Ranging System," Report No. PAT-
    APPL-6-425, 202, NASA-CASE-NPO-15865-1, 1982.
McPherson, C. A., "Three-Dimensional Robot Vision," *Proc. 3rd
    Int. Conf. Robot Vision Sensory Controls*, Cambridge, MA,
    Nov. 1983, pp. 167–178.
Meier, C., "Sensor Technology with Robot-Mounted Sensors," *Proc.
    4th Int. Conf. Robot Vision Sensory Controls*, London, Oct.
    1984, pp. 231–240.
Montagu, J. and Pelsue, K., "Ocular Manipulators for Robotic
    Vision," *Proc. 3rd Int. Conf. Robot Vision Sensory Controls*,
    Cambridge, MA, Nov. 1983, pp. 295–300.
Mori, S. and Tajima, S., "Applications of Robot with Vision Sen-
    sor," *Proc. '83 Int. Conf. Advanced Robotics*, Tokyo, Sept.
    1983, Part 1, pp. 367–374.
Morris, J., "The Use of Height Thresholding in Robotic Vision,"
    Proc. 2nd Eur. Conf. Automated Manufacturing, Birmingham,
    UK, May 1983, pp. 133–140.
Mortimer, G. and Newbury, L., "Robot Vision," *Pract. Electron.*,
    Vol. 19, No. 8, pp. 34–37.
Nakamura, Y. and Hanafusa, H., "A New Optical Proximity Sensor
    for Three Dimensional Autonomous Trajectory Control of Robot
    Manipulators," *Proc. '83 Int. Conf. Advanced Robotics*, Tokyo,
    Sept. 1983, pp. 179–186.
Nevskiy, I. V., Osadchiy, S. M., and Solntev, S. V., "Determination
    of Distances to Visible Points of Bodies in a System of Visual Per-
    ception," *Eng. Cybern.*, Vol. 19, No. 5, 1981, pp. 80–86.

Novak, A. and Colding, B., "Sensing of Workpiece Diameter, Vibration and Out-of Roundness by Laser—Way to Automate Quality Control," *Ann CIRP*, Vol. 30, No. 1, 1981, pp. 473–476.

Orrock, J. E., Garfunkel, J. H., and Owen, B. A., "An Integrated Vision Range Sensor," *Proc. 3rd Int. Conf. Robot Vision Sensory Controls*, Cambridge, MA, Nov. 1983, pp. 263–270.

Ozaki, N., et al., "Telé-Operated Robot for Inspection Inside of the PCV," *Trans. Am. Nucl. Soc.*, 1982 Winter Meet., Washington, DC, Vol. 43, Nov. 1982, pp. 738–739.

Page, N. S., Synder, W. E., and Rajala, S. A., "Turbine Blade Image Processing/Robot Vision System," *Proc. Advanced Software in Robotics*, Liege, Belgium, May 1983, pp. 7.C/1-10.

Perkins, W. A., "A Model-Based Vision System for Industrial Parts," *IEEE Computer Society, 3rd Int. Joint Conf. Pattern Recognition*, Nov. 1976, pp. 739–744.

Petersson, C. U., "An Integrated Robot Vision System for Industrial Use," *Proc. 3rd Int. Conf. Robot Vision Sensory Controls*, Cambridge, MA, Nov. 1983, pp. 241–248.

Petitto, R. M., "VUEBOT: A Fast All-Hardware Approach to Vision," *Electro '83 Electronics Show and Convention*, New York, April 1983, pp. 20/1/1-4.

Pinson, L. J., "Imaging Systems for Robotics Applications," *Proc. 3rd Int. Conf. Robot Vision Sensory Controls*, Cambridge, MA, Nov. 1983, pp. 279–284.

Pot, J., "Comparison of Five Methods for the Recognition of Industrial Parts," *Digital Syst. Ind. Autom.*, Vol. 1, No. 4, pp. 289–303.

Potter, R. D., "Applications of Industrial Robots with Visual Feedback," Soc. Mfg. Eng. Tech. Paper No. MS77-748, 1977.

Roy, A. and Sutro, L. L., "Simulation of Two Forms of Eye Motion and Its Possible Implication for the Automatic Recognition of Three-Dimensional Objects," *IEEE Trans. Syst. Man. Cybern.*, Vol. 12, No. 3, 1982, pp. 276–288.

"Stero Colour TV System for Industrial Application," *Neue Tech.*, Vol. 24, No. 10, 1982, pp. 40–41.

Tsuji, S., et al., "WIRESIGHT: Robot Vision for Determining Three-Dimensional Geometry of Flexible Wires," *Proc. '83 Int. Conf. Advanced Robotics*, Tokyo, Sept. 1983, pp. 133–138.

Vanderbrug, G. J., Albus, J. S., and Barkmeyer, E., "A Vision System for Real Time Control of Robots," *Proc. 9th Int. Symp. Industrial Robots*, March 1979, Washington, DC.

West, P. C., "Machine Vision in Practice," *IEEE Trans. Ind.*, Vol. 1A-19, No. 5, pp. 791–801.

Vision Systems

Agin, G. J., "Computer Vision Systems for Industrial Inspection and Assembly," *Computer*, May 1980, pp. 11-20.

Aleksander, I. (ed.), *Artificial Vision for Robots*, Kogan Page, London, 1983.

Aleksander, I., Stonham, T. J., and Wilkie, B. V., "Computer Vision Systems for Industry: WISARD and the Like," *Digital Syst. Ind. Autom.*, Vol. 1, No. 4, pp. 305-323.

Amat, J., Casals, A., and Clario, V., "Location of Workpieces and Guidance of Industrial Robots with a Vision System," *Proc. 4th Conf. Robot Vision Sensory Controls*, London, Oct. 1984, pp. 223-230.

Ambler, A. P., "Combining Vision Verification with a High Level Robot Programming Language," *Proc. 4th Int. Conf. Robot Vision Sensory Controls*, London, Oct. 1984, pp. 371-378.

Basanez, L. and Torras, C., "The Seeep Mapping in Robot Vision: Properties and Computational Savings," *Proc. 3rd IFAC/IFIP Symp.*, Madrid, Oct. 1983, pp. 127-132.

Batchelor, B. G., "Interactive Image Analysis as an Aid to System Design for Inspection and Robot Vision," *Proc. SPIE Int. Soc. Opt. Eng. (USA) Optical Sensing Techniques, Benefits Costs*, Teddington, UK, Vol. 376, Dec. 1982.

Bolles, R. C., "Three-Dimensional Locating of Industrial Parts," Final Report, Project SRI-1538, 1983.

Bracho, R., Schlag, J. F., and Sanderson, A. C., "POPEYE: A Gray-Level Vision System for Robotics Applications," Carnegie-Mellon University Report No. CMU-RI-TR-83-6, 1983.

Britanak, V., Sloboda, F., and Trebaticky, I., "A Modular Vision System for Inspection, Materials Handling and Assembly," *Proc. 3rd Scandinavian Conf. Image Analysis*, Copenhagen, July 1983, pp. 199-204.

Burgess, D. C., Hill, J. J., and Pugh, A., "Vision Processing for Robot Inspection and Assembly," *Proc. SPIE Int. Soc. Opt. Eng. (USA) Robotics Industrial Inspection*, San Diego, Vol. 360, Aug. 1982, pp. 272-279.

Chen, M. J., et al., "Integrated Vision-Based Workstations," *Proc. 14th Int. Symp. Industrial Robots/7th Int. Conf. on Industrial Robot Technol.*, Gothenburg, Sweden, Oct. 1984, pp. 587-598.

Christensen, C. P., Geesey, R. A., and Stickley, C. M., "Vision Systems for Intelligent Task Automation," *Proc. 4th Meet. Coordinating Group Modern Control Theory*, Rochester, MI, Oct. 1982.

"Faster Hardware, New Image-Processing Techniques Bring Robots
  Real-Time Vision," *Electronics*, Vol. 56, No. 23, 1983, pp.
  123–125.

Fu, K. S., "Pattern Recognition for Automatic Visual Inspection,"
  Computer, Dec. 1982, pp. 34–40.

Gleason, G. J. and Agin, G. T., "The Vision Module Sets Its Sight
  on Sensor-Controlled Manipulator and Inspection," *Robotics
  Today*, Winter 1980–81, pp. 36–40.

Gonzales, R. C. and Safabakhsh, R., "Computer Vision Techniques
  for Industrial Applications and Robot Control," *Computer*, Dec.
  1982, pp. 17–32.

Gorowara, H. K. and Brandeberry, J. E., "A Computer Vision System," *Proc. IEEE 1983 Nat. Aerospace Electronics Conf.*, Dayton,
  OH, May 1983, Vol. 1, pp. 207–209.

Goshorn, L. A., "Vision Systems Eye Real-Time Speeds through
  Multiprocessor Architectures," *Electronics*, Vol. 56, No. 25,
  1983, pp. 137–140.

Graham, D. and Choong, Y. C., "A Vision System for Surface Defects Detection Using an Object Model," *Int. Conf. Electronic
  Image Processing*, York, UK, July 1982, pp. 119–123.

Hunt, A. E. and Sanderson, A. C., "Vision-Based Predictive Robotic Tracking of a Moving Target," Carnegie-Mellon University
  Report No. CMU-RI-TR-82-15, 1982.

Jarvis, J. F., "Research Directions in Industrial Machine Vision: A
  Workship Summary," *Computer*, Vol. 15, No. 12, 1982, pp.
  58–61.

Jarvis, J. J., "Visual Inspection Automation," *Computer*, May 1980,
  pp. 32–38.

Jarvis, R. A., "Range from Brightness for Robotic Vision," *Proc.
  4th Int. Conf. Robot Vision Sensory Controls*, London, Oct.
  1984, pp. 165–172.

Karlsten, P., "ASEA Robot Vision Offers Increased Production Flexibility," *Proc. 7th Brit. Robot Assoc. Annu. Conf.*, Cambridge,
  UK, May 1984, pp. 165–170.

Kruger, R. P. and Thompson, W. B., "Technical and Economic
  Assessment of Computer Vision for Industrial Inspection and
  Robotic Assembly," *IEEE Proc.*, Vol. 69, No. 12, pp. 1524–1538.

Losty, J. A. and Watkins, P. R., "Computer Vision for Industrial
  Applications," *GEC J. Res.*, Vol. 1, No. 1, 1983, pp. 24–34.

Mandeville, J. R. and Meyer, J., "Robot Tool-Tacking Pattern Optimisation," *IBM Tech. Disclosure Bull.*, Vol. 26, No. 2, 1983,
  pp. 482–484.

Maruyama, T. and Uchiyama, T., "Real-Time Processor with Two
  Convolution Filter Modules and a Peak Extraction Module," *Proc.
  IEEE Computer Society Conf. Computer Vision Pattern Recogni-
  tion*, Washington, DC, June 1983, pp. 546–549.

Meagher, D. J., "Computer Software for Robotic Vision," *Proc. SPIE Int. Soc. Opt. Eng. (USA) Robotics Industrial Inspection,* San Diego, Aug. 1982, pp. 318–325.

Mortimer, J. M., "A Vision System for the Main on the Shopfloor," *Sensor Rev.,* Vol. 4, No. 1, 1984, pp. 39–42.

Movich, R. C., "Robotic Drilling and Riveting Using Computer Vision," *Robotics Today,* Winter 1980–81, pp. 20–29.

Myers, W., "Industry Begins to Use Visual Pattern Recognition," *Computer,* May 1980, pp. 21–31.

Nagel, R. N., Vanderburg, G. J., Albus, J. S., and Lowenfeld, E., "Experiments in Part A Acquisition Using Robot Vision," Society of Manufacturing Engineers, Paper No. MS79-784, 1979.

Park, D. C., "Development of a Simple Computer Vision System," *J. Korea Inst. Electron. Eng.,* Vol. 20, No. 1, 1983, pp. 1–6 (in Korean).

Pervin, E. and Webb, J. A., "Quaternions in Computer Vision and Robotics," *Proc. IEEE Computer Society Conf. Computer Vision Pattern Recognition,* Washington, DC, June 1983, pp. 382–383.

Porter, G. B. and Mundy, J. L., "Visual Inspection System Design," *Computer,* May 1980, pp. 40–49.

Rosenfield, A., "Computer Vision Research for Industrial Applications," *Proc. Trends Applications 1983, Automating Intelligent Behaviour, Applications Frontiers,* Gaithersburg, MD, May 1983, pp. 43–49.

Rossol, L., "Computer Vision in Industry—The Next Decade," General Motors Research Laboratory Publ. GMR-4096, 1982.

Rybak, V. I., "Investigation and Development of Visual Analysis and Environment Description Systems for Autonomous Manipulative Robots," *Cybernetics,* Vol. 18, No. 5, 1982, pp. 649–457.

Sanderson, A. C. and Weiss, L. E., "Image-Based Visual Servo Control of Robots," *Proc. SPIE Int. Soc. Opt. Eng. (USA) Robotics Industrial Inspection,* San Diego, Vol. 360, Aug. 1982, pp. 164–169.

Schachter, B. J., "A Matching Algorithm for Robot Vision," *Proc. IEEE Computer Society Conf. Computer Vision Pattern Recognition,* Washington, DC, June 1983, pp. 490–491.

Schaefer, D. H., Veronis, A. M., and Salland, J. C., "Spatially Parallel Architectures for Industrial Robot Vision," *Proc. IEEE Computer Soc. Conf. Computer Vision Pattern Recognition,* June 1983, Washington, DC, pp. 542–545.

Schreiber, R. R., "Robot Vision: An Eye to the Future," *Robotics Today,* Vol. 5, No. 3, 1983, pp. 53–57.

"Seeing Eye Robots for the Automated Factory," *Production Eng.,* Vol. 30, No. 8, pp. 48–51.

Selfridge, M., "Natural Language Interfaces to Image Analysis Systems," *Proc. Trends Applications 1983, Automating Intelligent Behaviour, Applications Frontiers*, Gaithersburg, MD, May 1983, pp. 248–251.

Shaheen, S. I. and Levine, M. D., "Some Experiments with the Interpretation Strategy of a Modular Computer Vision System," *Pattern Recognition*, Vol. 11, No. 6, pp. 87–100.

Siegler, A. and Bathor, M., "A Robot Vision System with Industrial Applications," *Proc. 7th Brit. Robot Assoc. Annu. Conf.*, Cambridge, UK, 1984, pp. 269–274.

Starks, N. V., "Robotics and Machine Vision in Electronic Production," *Electron. Prod.*, Vol. 12, No. 9, 1983, pp. 56–63.

Sugihara, K., "Deduction of Stero Structure from Line Drawings," *J. Soc. Instrum. Control Eng.*, Vol. 22, No. 2, pp. 201–208 (in Japanese).

Vanderburg, G. J., "Cybervision: A System for Flexible Robotic Assembly," *Electo/83 Conf. Record*, Boston, May 1982, pp. 23-1/1-1.

Vanderburg, G. J. and Nagel, R. N., "Image Pattern Recognition in Industrial Inspection," NBS Report No. 79–1764, 1979.

Villers, P., "Artificial Vision—Its Emergence as a Major Factor in the Robotics Industry," *Proc. 14th Int. Symp. Industrial Robots/7th Int. Conf. Industrial Robot Technol.*, 2-4, Oct. 1984, Gothenburg, Sweden, pp. 561–574.

Villers, P., "The Role of Vision in Industrial Robotic System and Inspections," *Electo '83 Electronics Show and Convention*, New York, April 1983, pp. 20/2/1-14.

Waldman, H., "Update on Research in Vision," *Robotics Today*, Vol. 5, No. 3, 1983, pp. 63–67.

Walters, R. A., "Robotic Vision Using Spatial Optical Sampling" (University of Central Florida, USA), in *IEEE Southeastern '83 Conf. Proc.*, Orlando, FL, April 1983, pp. 115–118.

Yachida, M. and Tsuji, S., "Industrial Computer Vision in Japan," *Computer*, May 1980, pp. 50–63.

9

Robot Applications

In this chapter we turn our attention to the applications of robots in manufacturing operations. First we will describe typical examples of applications in the following sectors: welding operations, automotive welding, machining operations, and finishing processes such as spraying of paint and other materials. Application of robots in production lines, forging, and heat treatment operations are described next. Flexible manufacturing systems utilizing robots are gaining popularity in many industrial operations. The final segment of this chapter is therefore devoted to a discussion of the features of flexible manufacturing systems. Many examples of robotic manufacturing cells are also described.

9.1 WELDING

Arc welding is one of the foremost processes used in industry for joining metals. The arc welding process requires dexterity and adaptive control. There are two levels of automated welding systems: fully automated and semiautomated. A fully automated system incorporates both position control and adaptive weld process parameter control. Some systems have add-on seam trackers for adaptive position control, which corrects deviations in the programmed weld path. In such a system, a position sensor just ahead of the weld gun monitors the position of the weld seam. The positional information so obtained is then used to drive the slides on which the weld gun is mounted. Adaptive process control is much more difficult since it requires the measurement of penetration, bead geometry, deposition rate, and weld quality in real time.

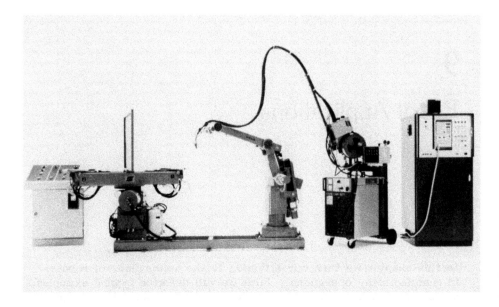

FIGURE 9.1   ESAB A30A robot arc welding system. (Courtesy of
ESAB Robotic Welding Division.)

Safety and productivity in arc welding dictate the use of robot
arc welding systems. The environmental conditions resulting from
the welding process, such as smoke, temperature, and radiation,
are potential hazards to the welder. Increased productivity in robotic arc welding is derived from the fact that the robot minimizes
nonproductive time in the welding cell. A robotic arc welding system permits better process control by the storage of predetermined
optimum weld parameters in the welding software of the robot controller. Consistent and high-quality welds are obtained by the repeatable positions and velocities of the robot. In robotic arc welding, the weld parameters are reproducible from part to part since
they are controlled by the robot's program.

The ESAB automatic robot arc welding system is shown in Figure 9.1. The nucleus of the welding station is the IR66 industrial
robot, which has rapid and precise movements that provide high-
capacity robotic production. For larger workpieces the IR60, with
approximately double the work envelope, may be used. By means
of the control panel, an operator determines when the robot should
start welding the next workpiece. The handling unit is an important

part of the welding station. The glare screen provides protection
for an operator clamping the next workpiece to the fixture while the
robot is welding on the preceding workpiece. Different types of
handling units are available for different jobs. On the type shown
in Figure 9.1, the fixtures are mounted on tiltable turntables for
accessibility. Several types of handling units are shown in Figure
2.21. The handling unit is often mounted on the same foundation
plate as the robot, producing constant stability and accuracy. For
very large workpieces welding stations with tracks could be used.
The welding feed unit and the wire reel travel with the robot on a
track, while the power source remains stationary. The arc welding
system includes a constant-voltage rectifier type solid-state power
supply, a wire feed unit, a welding gun with welding conduit assembly, and a welding programmer. There is a relieving arm for the

FIGURE 9.2 Arc welding of revolving frames. (Courtesy of ESAB
Robotic Welding Division.)

wire feed unit. The selectable welding programs include start/stop, gas preflushing, electrode feed, and nozzle flushing. Fast change-over from one welding job to another is made possible by programming storage on a tape cassette. Figure 9.2 shows an example of robotic welding.

Robotic welding has been shown proven to offer a threefold increase in productivity. Effective arc time in manual welding is 20 to 30%, whereas in robot welding it can be raised to 75% because the robot works quickly and continuously. All robot welds are identical and have uniformly high quality. It is easier to control the welding sequence to avoid problems with thermal stresses in the material. If any distortion occurs it is generally the same in every workpiece. Since robot programs can be rerun at any time, setup time is reduced with consequent speedup of the flow of work. Using two unit positioners with a separating glare screen enables loading of one table while the robot is welding on the other table.

9.2   AUTOMATIVE WELDING

Robots have been used for about 15 years to weld ocmplete automotive body assemblies and subassembly components (Figure 9.3). The early applications were mainly for off-line subassembly welding in fixed-position welding fixtures. The early welding guns were limited to 100 lbs (45 kg); later developments have expanded the applicability of robots in automotive welding with heavier guns and more dextrous wrist positions which can reach into areas with limited access. In robot welding installations performing spot welding on fixed, located body assemblies, the product must be stopped and located on a repetitive basis. This requirement is standard in either off-line subassembly or indexed body assembly systems. The problem with this mode of robot operation is that if the line stops or the weld gun sticks during production, the robot must be reprogrammed. Two other disadvantages are low-speed operation and poor repeatability.

Recent advances in this application are in moving line tracking. Tracking a moving line with a fixed-base robot is more costly and technically more difficult than the indexing line. Traversing-base robots eliminate the problems of base size and torque requirements of traversing floor-mounted robots. High arm strength with reduced floor requirements are achieved by using a cylindrical-type robot arm. For applications in which floor space is a problem, a compact welding robot may be mounted in an overhead configuration. Overhead mounting is advantageous in body respot lines and body side buildup areas with limited floor space.

FIGURE 9.3   Automotive body welding with Cincinnatti Milacron T³. (Courtesy of Cincinnatti Milacron.)

Robots have been used successfully in automation of truck cab welding. The welding line consists of nine 6-axis motion robots, allowing the welding guns to be positioned in any altitude. The cab fabrication area has six separate sections: rear sill subassembly, cab underbody sill subassembly, cab underbody line, side-panel assembly, door assembly, and the cab final assembly line.

In the underbody sill assembly line, two robots weld thick steel components (about 1/8 in. or 3.2 mm). In the next line, rear sill subassembly, a single robot spot welds the steel rear sills, which are fixtured on a two-station indexing table. The rear sill is finished by the robot spot welding anchor plates on the sill. Six more robots are used to complete the cab underbody spot welding. On the side-panel assembly four robots attend the lef - and right-side panels, which are delivered by a continuous overhead conveyer. The cab final assembly line has two robots which spot weld the cab front window opening, side panels, cab cowl, and cab roof inner panel and roof.

Most automotive assembly plants use a conventional gate line, which is prone to cause dimensional variations between identical body styles. In a novel approach, the underbody panel and two sides are locked up and tack welded in a large tooling fixture to establish a dimensionally sound body structure. The welding is done by eight robots. After this tooling fixture, the lightly tacked body shell moves to four other robots for pickup spot welding and then to a 24-robot respot line.

A robot welding station has been used to automate the production of rear axles. The robot welds mounting points for rear suspension, brake pipes, and stabilizers onto an axle tube in a very short cycle with a single torch. The robot maintains positional accuracy to within ± 0.008 in. (± 0.02 mm) on a completed rear axle assembly. Unlike the conventional method in a special-purpose multitorch welding station, the robot is easily reprogrammable to incorporate changes in axle design.

In another automobile plant a 6-axis industrial robot with tracking option is used to place spot welds on car bodies as they pass on an assembly line without stopping. This is accomplished by using position sensors which are mechanically linked to the conveyer line and electrically interfaced with the robot computer controls. The feedback signals are used by the robot control to continuously zero shift the previously taught program and compensate for movement of the car body. This enables the robot to place welds properly for any line velocity. If a line stops long enough that the robot cannot reach a spot, the robot waits until the line starts up and continues the welding where it left off.

In performing spot welding inside an automobile body, the robot can be mounted vertically on the garter which straddles the conveyer lines. A polar-type robot with a prismatic joint can then stretch its arm equipped with a spot welding gun into the roofless car body. In such a configuration the robot can weld the 44 points needed inside the car body with higher speed and accuracy than a human operator.

An automotive welding line contains 27 spherical-type robots. Traditional multispot welding jigs are eliminated at the second stations, where the bodies are tacked. Lineup and clamping are effected by a system of swinging gates at both sides and two from the top of the front and rear windscreens. These provide centering and clamping points and ensure correct positioning for the seven robots that perform the tack welds. Next is a pair of respot lines consisting of eight horizontal robots and two vertical robots. After the roof has been loaded automatically and precisely on the car body, a vertical robot makes tack welds and another spots in drip channel to fix the position. The tie plates are then loaded into an automatic transfer machine and held in position for spot welding by six robots. The final respot welding station has five robots. See Figures 9.4 and 9.5 for different automotive welding installations.

FIGURE 9.4   Automotive welding with KUKA IR 601/60 robots. (Courtesy of Expert Automation, Inc.)

FIGURE 9.5   Automotive welding installation.   (Courtesy of Expert
Automation, Inc.)

9.3   MACHINING OPERATIONS

Machining operations include deburring of machined parts, removal
of flash from parting lines of forged and cast parts, and drilling,
profiling, and riveting of aircraft panels. The tools used in these
operations are pneumatically or electrically driven. Some cases require adaptive control of the tool, and force feedback may also be
applied.

Burrs are often generated when metal parts are machined. Manual removal of burrs is a monotonous and expensive operation. In
special applications, sandblasting and explosive shocks may be employed in deburring. Using industrial robots is an economical method for automation of the deburring process. There are many different types of burrs, depending on how they are generated. Machining burrs are generated by drilling, milling, grinding, punching,
and cutting operations. Burrs from die-cast metal parts and some
molded plastic parts are similar to machining burrs. Excess materials from sand casting are often considerably heavier than machining
burrs. There are considerable variations in burr size, shape, and
brittleness. Robot deburring installation puts a high requirement
on the robot capabilities in contouring, repeatability, speed, servo
stability, and programming. The contouring capability is necessary
to follow an arbitrary contour with high accuracy to obtain satisfactory deburred quality. Very high repeatability is important in guar-
anteeing that the robot can move without noticeable variations. During a deburring operation different speeds are normally required for
different parts of the contour. Servo stability ensures that the
servo system will quickly compensate for varying loads as a result
of the different sizes of the burrs.

Deburring tools in robot applications are typically rotating high-
speed hard metal tools. In applications where it is necessary to
reach into sharp corners, reciprocating tools are ideal. Band sand-
ing machines may also be used. Support of the deburring tool is
also very important and must incorporate appropriately damped resilient mounting. Deburring quality and satisfactory cycle time are
obtained by ensuring correct contact point angle, path direction,
and velocity. The accessibility problem in deburring small holes
may be solved by using a rotating spherical file of small diameter.

In the automotive industry robots are used in deburring machined cast-iron brake regulators used in trucks and heavy vehicles.
The burrs are generated inside the detail at the intersections between holes drilled from different directions. In an installation with
a robot, the robot handles the part itself since the part is relatively
light in weight. Two deburring tools are used in this application:
a rotating hard metal tool and a rectangular reciprocating file, which
is used especially to reach the sharp corners on the contour. The

FIGURE 9.6   Polishing stainless steel sinks.   (Courtesy of ASEA, Inc.)

part is moved by the robot between the tools and manipulated in such a way that the tools follow the contours to be deburred very accurately and at the right speed.   As shown in Figure 9.6, a robot picks up stainless steel sinks and guides them against the buffing wheels, which are attached to fixed spindles.   This task is accomplished more rapidly and far more efficiently than if it were done manually.

9.4   FINISHING OPERATIONS

Finishing operations include the spray application of paints, stain, plastic powder, plastic sealer, sound deadener, or similar materials. Robots used in finishing operations are typically continuous-path, servo-controlled units having up to six or seven degrees of freedom. Hydraulically driven robots are intrinsically safe in the volatile atmosphere of such finishing operations.   Advantages of the use of robots in spray finishing include material and energy savings, consistent quality, reduced booth maintenance, and removal of operators from a potentially hazardous environment.

Manual spray painting is conventionally done by using air-stomizing spray guns. Newer methods is use are airless spraying, electrostatic spraying, and powder coating. Automatic machines parallel to these developments have been introduced to reduce the manual work, improve quality of spraying, and allow higher production rates. There is an overriding concern about solvent emissions, human health and safety, and fire hazards in paint operations. Robot spray painting, which is the ultimate in spray painting automation, eliminates the need for an operator to spend prolonged periods in the hazardous environment (Figure 9.7).

Painting robots are typically programmed by teaching a number of points on the required path with push buttons on the teaching box. The information on one point covers gun position and orientation, painting start and stop, painting speed, and other operational signals. The robot controller automatically moves the spray again according to the taught program. Features for synchronization with a moving conveyer enable a painting robot to paint a moving workpiece automatically as commanded by conveyer speed signals.

A painting robot follows the complex profiles of paints and reaches into cavities and recesses not possible with other forms of automation. A programmed spray sequence will be repeated and with reduced rejects and higher efficiency.

Most automotive spray painting is done with conventional horizontal and reciprocater-type spray machines, which are designed to paint large areas of the car. Some of the painting processes involve lower grill and headlight area, wheel wells, window openings, lower deck area, engine and truck compartments, and door facings. The robotic potential in this application is based on the robot's ability to work in the hazardous spray environment and allow the facility to operate with lower energy requirements. See Figures 9.8 and 9.9 for typical robot spray painting stations.

Spray painting is one of the most hazardous operations in the automotive industry that must be automated with a robot. In one installation, primer and finish color spray painting operations on urethane plastic bumpers have been automated with 16 servo-controlled robots. The new installation consists of four painting lines with four robots in each line.

The bumpers are conveyed on double-hung monorails in a loop. The robot is synchronized with the monorails by a pulse generator on each monorail, which feeds appropriate signals to the robots. Monorail start and stop automatically control the start and stop of the spray guns. Limit switches may be used as sensors for determining part presence and style of the bumper. The robots used have continuous path capability, which is necessary for precise

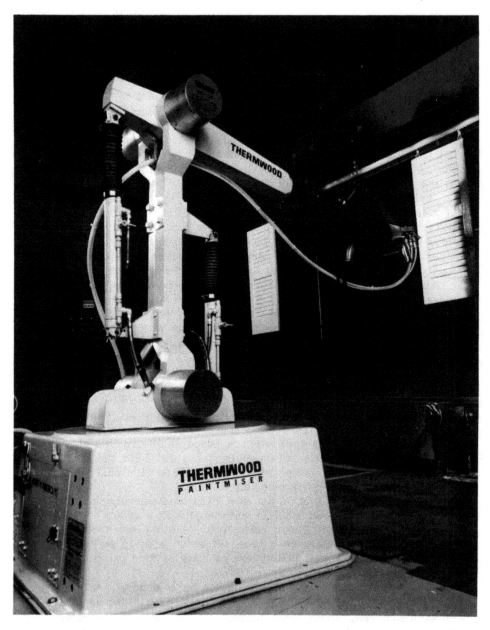

FIGURE 9.7   Spray painting with a robot.   (Courtesy of Thermwood Corporation.)

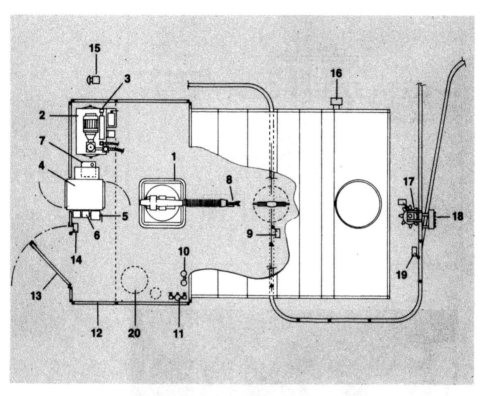

FIGURE 9.8   Layout of a robot spray painting station. 1) Manipulator, 2) hydraulic power supply, 3) water/oil heat exchanger,
4) control enclosure, floor standing (includes: a) robot controller,
b) hard disk memory, c) CRT terminal, d) air conditioner for enclosure.), 5) system disconnect, 6) transformers 220/110 or 440/110,
7) electrostatic power supply, 8) spray gun, 9) program start limit
switch, 10) gun trigger solenoid valves, 11) compressed air regulators and filter assembly, 12) wire mesh enclosure, 13) enclosure
gate, 14) gate interlock limit switch, 15) remote emergency stop PB
station, 16) exhaust flow switch, 17) memory parasitic drive assembly, 18) program identification manual input system, 19) program
"enter" limit switch, 20) material supply.   (Courtesy of Binks Manufacturing Company.)

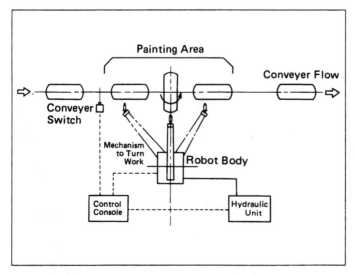

(a)

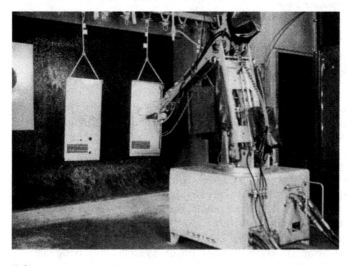

(b)

FIGURE 9.9   (a) Painting compressor air banks.   (b) Tokico painting robot installation.   (Courtesy of Tokico America, Inc.)

spray path motions. Paint savings in a robotized spraying are on the order of 10%, and there is a reduced requirement for ventilation compared to that for human operators.

In another installation two 6-axis robots spray the interior surfaces of pickup-truck boxes. The robots are programmed to spray the floor, tailgate, two sides, and front panel as the boxes move by at a line speed of about 12.5 ft/min (3.8 m/min). This installation provides for spraying four sizes of pickup-truck boxes in a variety of 22 colors.

To accomodate different models on an automobile spray line, several programs could be memorized and the appropriate one instantly selected when the robot receives the signal for a particular model. A speed dial enables easy alteration of painting speed to correct unsatisfactory paint quality. Universal painting robots permit the attachment of any painting gun and can be used for tasks other than painting, such as spraying anticorrosive oil and sealing chemicals.

In the ceramics industry, extremely unhealthy working conditions dictate the use of robots. It is difficult for operators to stand in the enamel while spraying the inside of a bathtub with enamel. In the sanitary industry, this type of application was one of the first jobs in which robots were used. Another common application is the glazing of washbasins and toilet tables. In coating cast-steel bathtubs, a robot shakes a mesh filled with powder enamel over the preheated surface to be coated. The bathtub must be turned over many times to have an even thickness of the film.

In the manufacture of microwave ovens, stainless steel cavities may be used. However, because of cost and consumer appeal, sheet metal boxes with excellent aesthetic appearance and corrosion protection are preferred. In a robot setup, the robot sprays a liquid baking enamel on the inside of the microwave oven cavities.

An example of an installation with a Spine spray system is a multirobot system fitted to a production line. Two types of vehicle chassis run on the production line. The production line passes through a spray booth 4.8 m wide. The control cabinet and hydraulic unit are located outside the spray booth.

Two robots can cover half the car chassis, producing a good result on both the inside and outside at a conveyer speed of up to 4.5 m/min. To cover the entire chassis, four robot systems are required. For speeds of 5 m/min or more it is necessary to have at least six robots, three on each side of the chassis. A recommended speed for the spray gun when spraying is 0.6 to 1.0 m/sec. Speeds up to 1.5 m/sec are used during the transport stroke.

Optical, pneumatic, inductive, or other types of sensors are sited on the production line. When the system is running under production line conditions, the control system uses the program

selection sensor to identify the object to be painted and then re-
trieve the relevant robot program. The conveyer belt and control
system are then synchronized with the help of the synchronizing
sensor. The synchronization point is the start point for the program. The position sensor then feeds a constant stream of information to the control system on the conveyer's speed and position.
With the conveyer in the correct position, calculated from the synchronization point, the program initiates an execution command and
the robot begins its task.

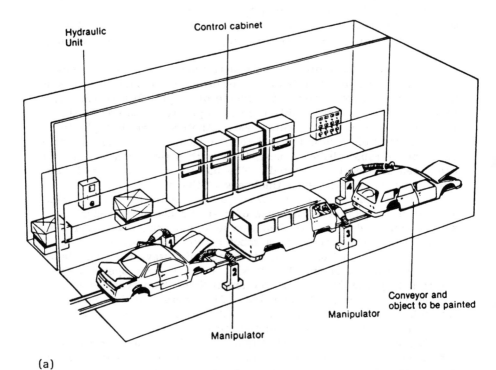

(a)

FIGURE 9.10   (a) Spine spray system. (b) Painting vehicle bodies
and details with a Spine robot. (c) Evaluation of different types of
integrated systems for totally automatic painting of car-body inte-
riors. (d) A Spine robot is well suited for painting long parts such
as the rear axle of an automobile. (e) Spine robots painting a van.
(f) Spray painting with the Spine robots. (Courtesy of Spine
Robotics Corp.)

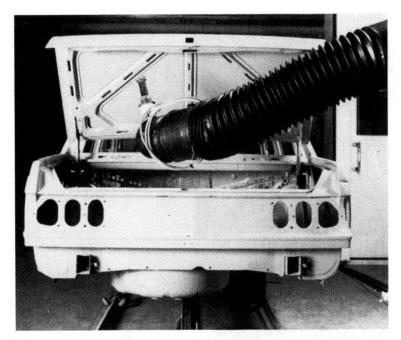

(b)

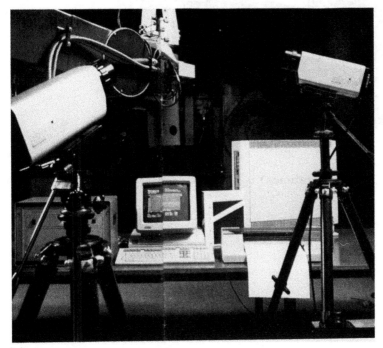

(c)

FIGURE 9.10   (Continued)

(d)

(e)

FIGURE 9.10   (Continued)

(f)

FIGURE 9.10   (Continued)

The process parameters, for example, spray on/off, paint flow
rate, and atomization pressure, are stored in the program and controlled by the computer. This system can be adapted to a specific
installation and, with the help of the process parameter functions,
can control completely tasks such as door opening, color change systems, or other peripheral systems. See Figure 9.10a-f for details
of the Spine spray system.

Gluing, puttying, and caulking constitute another application
area with potential for use of robots. In the automotive industry,
for instance, caulking of car bodies is a routine procedure. In
batch manufacturing lines, special fixed-program machines have been
used in the past. Robotized gluing becomes a practical alternative
in many applications. For example, in the assembly of heat ex-
changers or sealing of automobile engines, application of adhesive
is an essential stage in the manufacturing process. An industrial
robot applying adhesive to a car part is shown in Figure 9.11. The
system for this operation consists of (1) the servo-controlled robot,
(2) the hose arrangement, (3) the gun holder, (4) the application
gun and regulating equipment, and (5) the pump equipment that
feeds the adhesive to the dispensing gun. The upside-down configuration for the gluing process requires less space because the
robot is erected above the part and not beside it. Figure 9.12
shows a typical handling and sealing robot station.

FIGURE 9.11   Application of a string of adhesive to a car part.
(Courtesy of ASEA, Inc.)

FIGURE 9.12   Parts handling and sealant application of GCA robots.
(Courtesy of GCA Corporation.)

9.5   PRODUCTION LINES

A novel approach to robot technology utilizes both simple pick-and-place robots and flexible universal robots on the production lines of an automobile component factory. Three different production lines with 11 robot units are in operation at this factory: (1) shaft end production, (2) steering post assembly, and (3) main shaft production. These will be described briefly.

In the shaft end production line, a 5-axis universal robot picks up a rough forged blank from a conveyer and places it in a multi-purpose metal-cutting machine. The universal robot is appropriately suited to the advanced control patterns of the complex fixtures in the metal-cutting machine. The machined shaft end is unloaded by the robot and transferred to a simple pick-and-place robot. Loading and unloading of the lathe that machines the component is achieved by pick-and-place robots, which also make the final transfer to the outgoing conveyer leading to a deburring station and the final machining (thread cutting and drilling) center.

Production of the steering post assembly is in three stages. The steering mast is manufactured in one machining center and the main shaft is machined and assembled in a fully mechanized line attended by a robot. The robot unloads the lathe and transfers the machined steering masts to a special-purpose inspection and countersinking tool. The complex working pattern, difficulties in loading chucks, and workspace geometry require the use of universal robots in the main shaft production. The robot's first task is to load and unload the lathes in the first section of the machining line. The second task is to transfer the turned component from an intermediate station and load it into a drilling, rolling, and serration device. This station is the pickup position for a three-arm robot, which attends three other machines simultaneously: (1) the broach, (2) the slitting and calibration machines, and (3) the thread cutting and assembly unit.

Use of robots in this production line is motivated by high-quality production requirements, the need to reduce stockpiles of components and finished products, and labor shortage. These goals as well as flexibility and economic factors are being achieved by the application of both universal and pick-and-place robots.

Machining of automotive transmission cases is typically done by automatic transfer machines of the in-line lift-and-carry type. With the large number of stations, up to 170 per line, for these transfer machines, it has been necessary to break down each line into smaller sections with interlocking mechanisms manned for loading and unloading on demand. The automatic loading and unloading of transmission cases in the in-line lift-and-carry machine lines can be successfully implemented by using robots. An entire unit which has

been employed for a complete pallet transfer machine line includes
eight 6-axis robots, four for loading and four for unloading pallet
transfer machine sections. Each loading robot grasps the part, re-
moves it from the handling system carrier, and loads it into the
transfer machine pallet. The task of an unloading robot is to unload a pallet by grasping the part, lifting it through an external
blow-off cabinet, and finally transferring it to the handling system
carrier. Other schemes based on this application are possible in
automatic loading and unloading of transfer machines.

In another parts transfer application, an industrial robot is
equipped with a double gripper to load and unload steel connecting
rods in a duplex vertical broaching machine. The robot takes an
unfinished rod from the incoming conveyer, interchanges it with a
machined part in the broach fixture, and deposits the machined
part on an existing conveyer. The six-axis robot facilitates transfer in Cartesian motion with a fixed hand orientation. The straight-
line motion is especially useful in placing the pin end of the connecting rod on a tapered loactor in the broach fixture within a
clearance of 0.004 in. (0.10 mm).

In automobile bumper transfers, two bumpers are moved from a
stop-go conveyer on which they are built to an overhead monorail
conveyer. The overhead conveyer stops approximately in sequence
with the build conveyer, but its stationary position may vary widely
from one cycle to the next. Tracking capability is used to monitor
the position of the monorail conveyer for correct positioning of the
bumpers in the cradle. Heavy-duty robots with tracking capability
have also been used in automobile body-side transfer applications.
Body sides are picked up from a continuously moving monorail conveyer. This application requires a robot system with the ability to
track two continuously moving independent lines.

Manufacture of diesel engine parts by manual systems is a boring and repetitive job. Hard automation is very costly and special-
purpose automation does not have sufficient flexibility for product
changes. Robots have been used advantageously to improve the
manufacture of diesel engine parts. Two of the parts produced are
cam follower levers and cam follower housing. In the cam follower
lever line, levers in sets of three are loaded between two milling
machines by a robot. Milled sets are unloaded into a transfer conveyer, where a second robot picks lever sets and loads them into
machining transfer line. The basic operation in the cam follower
housing line is for the robot to load and unload drilling and milling
machines.

Proper dimensional control of the completed body of an automobile is vital for good door, hood, deck, glass, and molding fits.
Quality control checks of the body shell are at points located in and
around the windshield, door, and back window openings. Four

robots with repeatability on the order of ±0.005 in (±0.13 mm) and accuracy of ±0.010 in (±0.25 mm) have been used to replace manual body checking. Each robot has three probes mounted in different positions so that they can reach each programmed check point with a minimum or program steps. The probe toolholder is configured to facilitate making all the physical moves required for checking.

Robot shifters for manual transmission testing are operating in a computerized dynamometer testing facility. A greater number of cars can be tested simultaneously with the special robot shifters than with manual operations. The robot is actually a series of metal struts and pneumatic actuators positioned on the floor between the driver's seat and the controls. The installation consists of one actuator attached to the clutch pedal and two to the gearshifts for side-to-side and fore-and-aft shifts.

9.6   FORGING AND HEAT TREATING

Forging and heat treating are operations in which unpleasant conditions such as heat, noise, dirt, and heavy loads may be present. In these operations, special attention is necessary to protect the robot from heat and shock loads. Cooling of the robot's hand tooling may also be required. Robot applications in forging, upset forging, and roll forging involve operations in which robots may load furnaces, forging presses, headers, and trim presses.

In head treating tasks, robots may load and unload furnaces and quench parts. Robots have been used in forging operations to perform the following tasks:

Feed billets through two-cavity die forging press to produce raw
    differential side gears.
Transfer hot (2000 to 2200°F) diesel engine crankshafts from forging
    press output nest into a journal twister.
Transfer hot hand tool component platters from an impact forging
    system to hot trimming presses.
Load, unload, and lubricate aluminum closed die forge presses.
Precision forge jet engine airfoils.
Load harrow discs from pallets into a heating furnace in an aus-
    forming line.

Plastic molding operations which involve the use of robots are injection molding, structural foam molding, and compression molding. When robots are equipped with appropriate hand tooling, they can load and unload multiple cavity molds in a single operation. Also, robots can reach into large machines which are dangerous for human operators. Robots used in plastic molding operations perform the following typical functions

Load inserts and compression molding charges.
Unload injection molding machines.
Palletize and package moldings.
Trim moldings on removal from machine.

In die casting, the primary use of robots involves unloading die casting machines and handling the castings and scrap. The robots may perform tasks such as:

Remove casting from machine and place it on a conveyer or into a
    container.
Unload the casting, quench it, and place it in a trimming press.
Dispose of sprue and runners after trimming.
Load inserts into the die.

In the investment casting process, robots are used in the manufacture of shell molds for careful and consistent handling of the shells. Applications include:

Making ceramic shell molds for leisure-time products ranging from
    outboard motors to snowmobiles
Mechanical dipping and transfer to conveyer
Ceramic molds for blades, vanes, and some structural components
    used in airborne and land-based gas turbine engines

In foundry operations, robots are used to in a variety of applications such as:

Unload core-making machines.
Handle cores through mashing operations and set cores in molds.
Dry and vent molds.
Clamp and unclamp molds on pouring lines.
Transfer hot castings from molding lines to shake out machines.
Remove gates, risers, and sprues, and flash from castings.

Other examples illustrating the use of robots for material handling and machine loading are the following:

Load air-conditioning compressor housings into multiple horizontal
    turning machines.
Load connecting rod bearing caps in a large horizontal broaching
    machine.

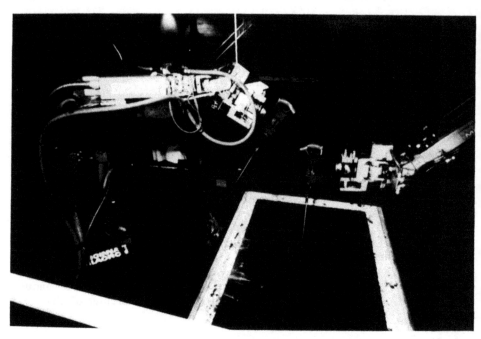

(a)

(b)

Unload large parts from injection molding machines making elastomer
   rubber parts.
Load and unload an automatic chucking machine and a broaching
   machine with differential case castings in a machining center.
Load abrasive machines in the manufacture of plugs for tapered
   plug-type valves.
Load and unload automatic glass edge grinding machines for automotive flat window glass.
Load and unload heat treat furnaces.
Load thermometer boxes into centrifuge and hot-water bath in the
   manufacture of clinical thermometers.

Figure 9.13 shows different uses of robots for handling and loading.

9.7   FLEXIBLE MANUFACTURING SYSTEMS

A flexible manufacturing system is a production facility which consists basically of a set of workstations for performing a number of
different tasks. A variety of workpieces are transported automatically under computer control from one machine to another. The
only manual interventions are performed at the loading and unloading stations located at the entry to the flexible manufacturing
system.
    The objective of such a system is to achieve the flexibility of
conventional or stand-alone machinery and maintain the higher productivity of a dedicated transfer line. Dedicated machinery is used
for production of at least 10,000 identical parts. Conventional
stand-alone machinery has very low production levels at the expense
of any flexibility and adaptation.
    Flexibility is the most important aspect of a flexible manufac-
turing system. The major characteristics of flexibility are:

1. Mix flexibility: processing at any one time a mix of different
   parts belonging to the same family.
2. Parts flexibility: parts can be added to and removed from the
   mix over a period of time.

FIGURE 9.13   Applications of a robot. (a) Putting vent holes in
casting molds requires controlled-path straight line motion features
in programming. (b) Loading and unloading forging presses and
injection molding machines can conveniently be done by a robot.
(Courtesy of Cincinnatti Milacron.)

3. Routing flexibility: the dynamic assignment of machines; a part can be rerouted through the system if a machine used in its manufacture is incapacitated.
4. Design change flexibility: fast implementation of engineering design changes for a particular part.

Ideally, a flexible manufacturing system should have all these properties. In practice, however, cost factors limit the flexibility of manufacturing systems.

Flexible manufacturing systems incorporate many individual automation concepts into a single production system. These include numerical control of machine tools and computer numerical control (CNC), direct numerical control (DNC), automatic materials handling machines, and group technology principles.

Direct Numerical Control

Direct numerical control means connecting a set of numerically controlled machines to a common memory for part program or machine program storage. It also has provision for on-demand distribution of data to machines. There is not a clear distinction between DNC and flexible manufacturing systems (FMS). Some writers use the two terms interchangeably, whereas others use DNC to describe a system where the central computer supplies part programs only to NC machines. Three types of DNC systems may be considered. The first is individually controlled DNC, in which a number of machine tools are run independently by a control computer.

The second type of DNC is the tandem type, which is used for parts having similar shapes and machining processes. In this case a series of NC machine tools are laid out in tandem according to the machining process and are connected by an automated conveyer. The system may also have automatic loading and unloading devices so that the entire process is controlled by a computer. Work in progress buffers may also be permitted. This type of DNC is highly productive, but one machine breakdown may damage the entire system. Although this system has some characteristics of an FMS, it is limited somewhat because of the fixed sequence of operations and forward movement of the conveyer. Because this system resembles a transfer line, it is sometimes called a DNC line.

The third type of DNC is the flexible type where a number of NC machine tools and automated materials handling systems are connected to a control computer. Workpieces are set up on pallets for random transfer, identification, and final positioning in sequence on the right machine tool. An automated storage or warehousing facility and automatic tool and jig exchange may also be used in this system. This is often referred to as the flexible manufacturing system.

Elements of Flexible Manufacturing Systems

The major equipment in an FMS falls into three main categories:

1. Computer hardware
2. Machine tools
3. Material handling system

Typical functions performed by a computer in an FMS are

Computer control at the individual machines (CMC)
Direct numerical control of all the machines in the FMS (DNC)
Control of material flow through the system
Monitoring of tool usage and machine utilization
Schedule of machining operations
Handling of part programs
Generation of management information
Performance of system diagnostics

Machine tools in an FMS produce either prismatic or rotational parts. For prismatic parts horizontal or vertical boring and milling machines may be used. Rotational parts require the use of turret and long-bed lathes. Special features of machine tools required are automatic tool and workpiece changing as well as tool wear and breakage detection. Material handling systems in FMS perform material handling both between machines or cells and at the machine or cell level.

Material handling systems between machines or cells within an FMS are of two main types: random access and addressable delivery. In random access systems, a pallet with parts circulating in a closed loop is pulled off the loop when it reaches a machine tool as required. Roller conveyers are most commonly used for random access material handling. Another system which may be used is two carts. A continually moving chain under the floor tows a trolley carrying the pallets through a shot in the floor. At a desired station, the link to the chain is disconnected to unload the part. When the link is reestablished the trolley continues to move.

In material handling with addressable delivery, a system computer fully controls carts which can move in any direction along a track. Addressable delivery systems can handle more than one pallet on a cart. Also, the movements of the cart can be chained for minimum time delays. These features make addressable systems more flexible in transporting parts from one station to another.

Design requirements for intermachine material handling are:

1. Random, independent measurement of palletized workparts between workstations in the FMS. This requires that the pallets be coded and that there be associated decoder switches at individual workstations to recognize pallets.

2. Temporary storage or banking of workparts. This can be imple-
    mented by using a central pallet magazine or individual cell
    magazine.
3. Convenient access for loading and unloading workparts.
4. Compatibility with computer control.

The functions of a flexible machining complex may be summarized
as follows:

Machining: turning, grinding, gear cutting, laser treatment, and
    adaptive control
Material handling: direct and indirect material handling
Auxiliary functions: diagnostics for failure detection, accuracy
    compensation, chip control, spindle cooling

A flexible assembly has the following functions:

Assembly: thread fastening, pressing and insertion, holding and
    complex control
Buffering: stacker for parts, jigs, and fixtures
Auxiliary functions: parts replacement, inspection, jig and fixture
    changing, and detection of parts

A flexible product inspection complex has the following functions:

Inspection: size, shape, vibration, noise, temperature
Material handling: pallet transportation
Auxiliary functions: preparatory work and repalletizing

Flexible manufacturing systems may also be divided into two
types: dedicated FMS and random FMS. The dedicated FMS is de-
singed to fulfill well-defined manufacturing requirements over a
period of time. It is similar to the tandem-type DNC or the DNC
line. The random FMS is designed to handle a greater variety of
parts in random sequence. This system corresponds closely to the
flexible type DNC. Within the concepts of FMS, a distinction may
be made between flexible manufacturing cell (FMC), flexible manufacturing module (FMM), and FMS.
    In a flowline design manufacturing system, the machine tools
are grouped according to some well-defined combination of process
and capacity requirements. The goal of a flowline system is the
manufacture of a complete family of parts. Many flowline systems
with robots are beginning to appear in manufacturing industries.
    A typical European flowline system consists of a turning center,
a rotary table surface grinder, two turret drills, a flip-over station, a parts conveyer, and a robot. This manufacturing cell makes

three different parts and six sizes. Parts of one size are machined
at each setup to minimize production and delivery problems and cost
of flexible tooling. In another installation an experimental flowline
system consists of a vertical spindle CNC milling machine, an NC
turret drilling machine, a horizontal spindle machinery center, a
CNC lathe, and an anthropomorphic robot. This system is being
used to determine the maximum level of automation that can be
achieved along with maximum flexibility.

The components of a manufacturing cell are

Processing: workstation manufacturing and assembly
Material flow: handling and storage
Human-machine interaction: uses programming and operation
    scheduling
Information flow: setup planning and process supervision

A functional manufacturing layout uses machines in specific
groups determined by the type of machining process. A stand-
alone functional manufacturing system has been developed by Cincinnatti Milacron (Figure 9.14a-c). This system, called the CIN-
TURN/T^3 Industrial Robot Manufacturing Cell, consists of

Two 12-in. universal turning centers
Gaging station
Robot system for presenting raw stocks and removing finished parts.

In a typical cycle of the manufacturing cell, the robot moves to the
conveyer carrying parts in random order and signals the conveyer
to start. A limit switch typically sends a signal when a part is
present in the pickup station. The part in the gripper is identified
by the internal memory of the robot controller. The robot then
moves to the available turning center. At this center, the control
sends appropriate input signals to check that the machining cycle is
completed and the door is open. The robot then enters the work
area to remove the finished part and lower the raw stock into the
holding device. The double gripper is useful for reducing the load
and unload time to a minimum.

The robot transfers the machined spindle to the gaging station,
where the part is set in the path of a laser beam for inspection.
The laser system immediately makes necessary tool adjustments and
displays the outer diameter dimensions on the readout. After the
laser inspection, if the part is within specified tolerances it can be
placed in a bin, palletized, or placed on another conveyer.

Automatic mounting of car wheels by means of industrial robots
is shown in Figure 9.15. In the pilot system, Figure 9.16, the robots pick up the heavy wheels and convey them to the correct

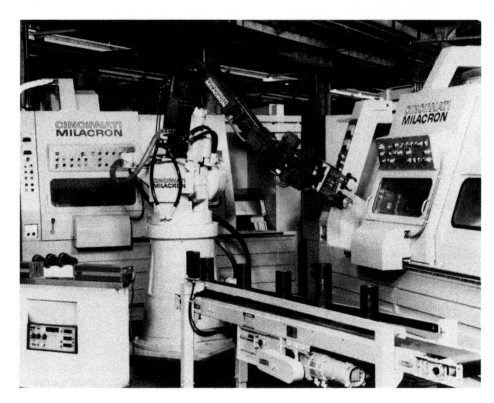

(a)

FIGURE 9.14   A functional manufacturing system.   (a) Cinturn/T^3
robot manufacturing cell.   (b) Spindle is held between center driver
and heavy duty tailstock.   (c) Robot places finished spindle in path
of laser beam to simultaneously update tool comps, if necessary, and
display OD dimension on gage readout.   (Courtesy of Cincinnati
Milacron.)

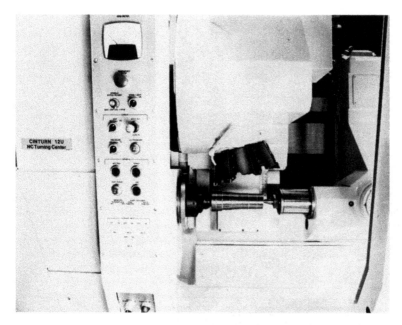

(b)

(c)

FIGURE 9.14   (Continued)

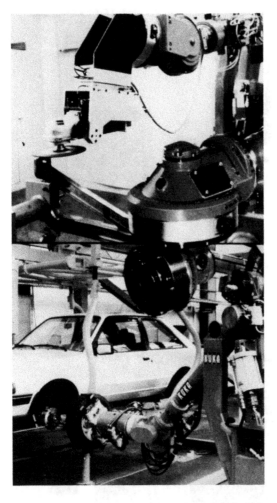

FIGURE 9.15   (Top) Robot with continuous-path clean castings,
(bottom) robot fits wheels to hubs. (Courtesy of Expert Automation, Inc.)

position, where they are swiftly and reliably bolted by robot to the
hub. The operation requires visual sensors to synchronize the vehicle and robot movements. The wrist of the robot is equipped with
a wheel gripper and multiple wrench. A television camera detects
the hub position and hole pattern. Acting on corresponding control
signals, the robot positions the wheel on the hub and bolts it tight.
    Spot welding of left and right car wheel housings provides an
example of flexible manufacturing system. In the facility shown in
Figure 9.17, left and right car wheel housings are being welded

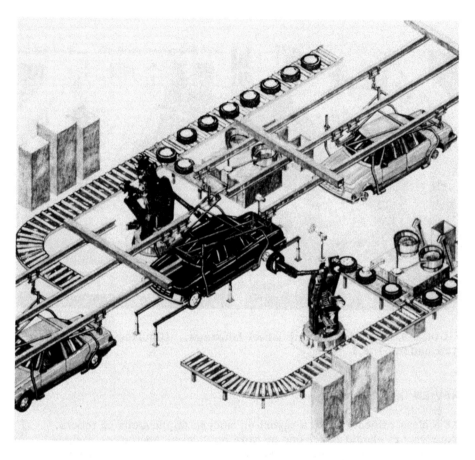

FIGURE 9.16   Wheel fitment system.   (Courtesy of Expert Automation, Inc.)

from the outside and the inside.   This requires a turntable with
four stations equipped with different types of backing electrodes.
The wheel housings, which are clamped to the backing electrodes,
are pivoted to the respective working zones of two portal robots by
rotating the turntable alternately to the left and the right by 90°.
When the appropriate program is activated by the external control,
the portal robots approach each point automatically and weld with the
optimum parameters for the respective job.   To cater for a random
model mix or change of type, only the backing electrodes must be
exchanged for other ones.

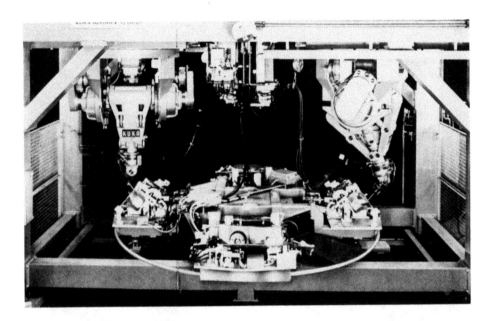

FIGURE 9.17   Spot welding wheel housings.   (Courtesy of Expert Automation, Inc.)

REVIEW QUESTIONS

As a class project, write a report on modern applications of robots. Your report should cover one or more application sectors as assigned by the instructor. Suggested areas to consider include automobile industry, manufacturing industry, materials handling, flexible manufacturing systems, finishing operations, assembly, heat treatment, aerospace, intelligent robots, nuclear, space exploration, casting and molding, and medicine.

FURTHER READING

Baylou, P., et al., "Agricultural Robotics," *Nouv. Autom.*, Vol. 27, No. 28, 1982, pp. 54–63.
Beni, G. and Hackwood, S., "Robotic Electroplating of Gold on Quaternary Semiconductors," *J. Appl. Electrochem.*, Vol. 13, No. 4, 1983, pp. 531–533.
Cronshaw, A. J., "Automatic Chocolate Decoration by Robot Vision," *Proc. 12th Int. Symp. Industrial Robots/6th Int. Conf. Industrial Robot Technol.*, Paris, June 1982, pp. 249–257.

DiPietro, F. A., "Organization of Assembling Cells by Robotics,"
    Proc. 8th Int. Conf. Industrial Robots, Detroit, June 1984,
    Vol. 1, pp. 2.39–2.62.
Drozda, T. J., "Robotized 'Multicells' Best for Mid-Volume Output,"
    Robot. Today, Vol. 5, No. 2, 1983, pp. 53–56.
"Electronic Gauges Keep Missiles within Limits in Robot Cell," Prod.
    Eng., Vol. 61, No. 6, 1982, pp. 16–18.
Ermert, W. and Menges, G., "Manufacturing of Components from
    GRP Using Industrial Robots," Plastverarbeiter, Vol. 33, No. 4,
    1982, pp. 399–403.
"Exploiting the Capabilities of Modern Adhesives," Industrial Robot,
    Vol. 10, No. 2, pp. 132.
Fleck, J., "Systemic Use of Robots," Data Processing, Vol. 25, No.
    12, 1983, pp. 19–23.
Freund, E., Hoyer, H., and Mehner, F., "Multi-Robot Systems in
    FMS," Proc. 3rd Int. Conf. Flexible Manufacturing Systems and
    17th Annu. IPA Conf., Boeblingen, West Germany, Sept. 1984,
    pp. 207–226.
Hess, K., "Variable Automatic Mounting of Machine Subassemblies by
    Assembly Robots in Mounting Cells," Fertigungstech. Betr.,
    Vol. 33, No. 7, 1983, pp. 399–402.
Houtzeel, A. and Chevalier, P., "Group Technology/Robotics,"
    Electro/82 Conf. Record, Boston, May 1982, pp. 23–4/1–5.
Irwin, R. D., "Are Robots Allergic to Rubber?", IEEE Conf. Record
    1983 35th Annu. Conf. Electrical Eng. Problems Rubber Plastics
    Industry, Akron, OH, April 1983, pp. 39–42.
Johnson, T. A., "Robot Controlled Laser Processing," Proc. 1st Int.
    Conf. Lasers Manufacturing, Brighton, UK, Nov. 1983, pp.
    71–80.
Kemp, D. R., et al., "A Sensory Gripper for Handling Textiles,"
    Proc. 13th Int. Symp. Industrial Robots and Robots 7, Chicago,
    Vol. 2, April 1983, pp. 18.23–18.33.
Kennedy, J. P., "Using Process Computers as Robots for Refining
    Operations," Hydrocarbon Process., Vol. 61, No. 5, 1982, pp.
    160–164.
Kochan, A., "FMS—An International Overview of Applications,"
    Proc. 1st Int. Machine Tool Conf., Birmingham, UK, June 1984,
    pp. 73–80.
Kusiak, A., "Robot Applications in Flexible Manufacturing Systems,"
    Proc. 8th Int. Conf. Industrial Robots, Detroit, June 1984,
    Vol. 1, pp. 3.1–3.13.
Maiette, R. L., "How to Plan a Robotic Machining Cell," Tool. Prod.,
    Vol. 49, No. 4, 1983, pp. 83–87.
Medeiros, D., "Simulation of Robotic Manufacturing Cells: A Modu-
    lar Approach," Simulation, Vol. 40, No. 1, 1983, pp. 3–12.
Mercer, A. and Wong, J., "Serial Interface Stars in Robotic Sta-
    tion," Electron. Des., Vol. 31, No. 10, 1983, pp. 177–182.

Moldenhauer, H. G. and Klaeger, W., "State and Trends in Development, Production, and Utilization of Industrial Robots in Heavy-Machine and Plant Building," *Fertigungstech, Betr.*, Vol. 33, No. 7, 1983, pp. 411–413.

Mollo, M., "A Distributed Control Architecture for a Flexible Manufacturing System," *Proc. 3rd Int. Conf. Flexible Manufacturing Systems* and *17th Annu. IPA Conf.*, Boeblingen, West Germany, 1984, pp. 227–240.

Mudge, T. N., "Hardware/Software Transparency in Robotics through Object Level Design," *Proc. SPIE Int. Soc. Opt. Eng. Robotics Industrial Inspection*, San Deigo, Vol. 360, Aug. 1982, pp. 216–223.

Ostapchuck, V. G., "Use of Adaptive Industrial Robots for Certain Mechanical Engineering Operations," *Sov. Eng. Res.*, Vol. 2, No. 5, 1982, pp. 3–6.

Page, G. J., "Vision Driven Stack Picking in an FMS Cell," *Proc. 4th Int. Conf. Robot Vision Sensory Controls*, London, Oct. 1984, pp. 1–12.

Partington, G., "Glass Gathering by Robot," *Glass Technol.*, 1982, Vol. 23, No. 2, pp. 80–89.

Puente, E. A., et al., "DISAM/2—An Experience in Flexible Manufacturing Systems," *Proc. 3rd Int. Conf. Flexible Manufacturing Systems* and *17th Annu. IPA Conf.*, Boeblingen, West Germany, 1984, pp. 311–322.

Raju, V. and Karlson, T. J., "Flexible Manufacturing Cells with Robots," *Proc. 3rd Int. Conf. Flexible Manufacturing Systems* and *17th Annu. IPA Conf.*, Boeblingen, West Germany, 1984, pp. 165–174.

"Robot Operates Unmanned Manufacturing Cell," *Prod. Eng.*, Vol. 61, No. 7, 1982, pp. 33–35.

"Robot Work Cell Rebuilds Milling Tools," *Prod. Eng.*, Vol. 30, No. 5, 1983, pp. 14–15.

Schneider, R. and Ziemann, G., "Introduction of the Industrial Robots Techniques in the Department of Vehicle Maintenance of the DR," *Eisenbahntechnik*, Vol. 31, No. 5, 1983, pp. 183–184.

Schreiner, L., "7-Axis Robot Automates Manufacturing Work Cell," *Electron. Des.*, Vol. 31, No. 23, 1983, pp. 141–148.

Stephens, O., "'Industrial Robot' Moves in on Lens Edging," *Ind. Diamond Rev.*, Vol. 42, No. 493, 1982, pp. 356–359.

Urbaniak, D. F., "The Unattended Factory Fanuc's New Flexibility Automated Manufacturing Plant Using Industrial Robots," *Proc. 13th Int. Symp. Industrial Robots* and *Robots 7*, Chicago, April 1983, Vol. 1, pp. 1.18–1.24.

Vuorilehto, J., "Production of Heavy Prismatic Parts in a Flexible
    Manufacturing System," *Proc. 14th Int. Symp. Industrial Robots/
    7th Int. Conf. Industrial Robot Technology*, Gothenburg,
    Sweden, Oct. 1984, pp. 159–166.

Warnecke, H. J., "Flexible Assembly System for Unmanned Factory,"
    Proc. 4th Int. Conf. Assembly Automation, Tokyo, 1983, pp.
    116–122.

"Why We Are Enthusiastic about Industrial Robots," *Mech. Eng.*,
    Vol. 38, No. 13, 1983, pp. 54–57.

Wright, P. K., "A Flexible Manufacturing Cell for Swaging," *Mech.
    Eng.*, Vol. 104, No. 10, 1982, pp. 76–83.

General Applications

Balsmeir, P. W., "Robot Systems for the Factory," *Arkansas Bus.
    Econ. Rev.*, Vol. 16, No. 1, 1983, pp. 1–9.

Bejozy, A. K., "Remote Applications of Robots," *Proc. '83 Int.
    Conf. Advanced Robotics*, Sept. 1983, Pt. 2, pp. 1–12.

Billingsley, J., Naghy, F., and Harrison, D., "The Craftsman
    Robot," *Electron. Power*, Vol. 29, No. 11, 1983, pp. 850–808.

Caplan, N., "Robots Are Coming, the Robots Are Coming," *IEEE
    Control Syst. Mag.*, Vol. 2, No. 2, 1982, pp. 10–13.

Davenport, D. A., "The Industrial Robot," *1983 Rochester Forth
    Applications Conf.*, Rochester, N.Y., June 1982, pp. 1–8.

Finlay, P. A., "Industrial Robots Today—A Brief Overview," *Mater.
    Des.*, Vol. 4, No. 3, 1983, pp. 776–782.

Gorman, R., "What to Look for in Industrial Robots," *Tool Prod.*,
    Vol. 49, No. 5, 1983, pp. 49–51.

"Industrial Robot," *Mod. Mater. Handl.*, Vol. 38, No. 4, 1983,
    pp. 46–53.

"Industrial Robots of the World," *Industrial Robot*, Vol. 10, Nos.
    1/2, pp. 135–137.

Kamm, L. J., "Recent Applications of Modular Technology Robots,"
    Proc. 13th Int. Symp. Industrial Robots and *Robots 7*, April
    17–21, 1983, Chicago, Vol. 1, pp. 11.66–11.75.

Keller, E. L., "Clever Robots Set to Enter Industry En Masse,"
    Electronics, Vol. 56, No. 23, 1983, pp. 116–119.

Miller, R. K., "Machine Vision for Robotics and Automated Inspec-
    tion," *Tech. Insights*, Vol. 2, 1983, pp. 108.

Morris, H. M., "Where Do Robots Fit in Industrial Control?", *Con-
    trol Eng.*, Vol. 29, No. 3, 1982, pp. 58–64.

Ottinger, L. V., "Robotics for the IE—The Automated Factory,"
    Ind. Eng., Vol. 14, No. 9, 1982, pp. 26–32.

Prono, M., "Adaptable to Every Demand (Potential Uses of Industrial
    Robots)," *Ind.-Anz.*, Vol. 105, No. 17, 1983, pp. 58–59 (in
    German).

Aerospace Industry

Akin, D. L., et al. "Space Applications of Automation, Robotics, and Machine Intelligence Systems, (ARAMIS) Phase 2," NASA Report NAS 1.26: 3734-6, 1983.

Fernandez, K., et al. "NASA's Use of Robotics in Building the Space Shuttle," Soc. Manuf. Eng. Tech. Paper Ser. MS 83-364.

Fernandez, K., Jones, C. S., III, and Robots, M. L., "NASA's Use of Robotics in Building the Space Shuttle," *Proc. 13th Int. Symp. Industrial Robots* and *Robots 7*, Chicago, April 1983, Vol. 1, pp. 11.35-11.43.

Johnson, D. G. and Hill, J. J., "A Sensory Gripper for Composite Handling," *Proc. 4th Int. Conf. Robot Vision Sensory Controls*, London, Oct. 1984, pp. 409-416.

Lambeth, D. M., "Robotic Fastener Installation in Aerospace Sub-assembly," *Proc. 13th Int. Symp. Industrial Robots* and *Robots 7*, Chicago, April 1983, Vol. 1, pp. 10.1-10.12.

Michaud, L. L., "Robots in Aerospace: A Multi-Tasking Robot for the Manufacture of Precision Guided Missile (PGM) Structures," *Proc. 13th Int. Symp. Industrial Robots* and *Robots 7*, Chicago, April 1983, Vol. 1, pp. 11.14-11.34.

Molander, T., "Machine of Sheet Metal Plates in Aircraft Manufacturing—Why Use an Industrial Robot?", *Robots 8, Proc. 8th Int. Conf. Industrial Robots*, Detroit, June 1984, pp. 12.1-12.10.

Movich, R. C., "An Intelligent Robotic System for Aircraft Manufacturing," *Robots 8, Proc. 8th Int. Conf. Industrial Robots*, Detroit, June 1984, pp. 12.11-12.29.

Scoville, J. H., "Robotic Abrasive Process System," *Robots 8, Proc. 8th Int. Conf. Industrial Robots*, Detroit, June 1984, Vol. 1, pp. 5.52-5.61.

Sullivan, M. J., "The Use of Industrial Robots in the Layup of Composite Material for Aircraft and Other Products," *Robots 8, Proc. 8th. Int. Conf. Industrial Robots*, Detroit, June 1984, Vol. 1, pp. 12.30-12.40.

Assembly

Albo, R. T., "Robot Assembly of Floppy Disk Drives," *Proc. 13th Int. Symp. Industrial Robots* and *Robots 7*, Chicago, Vol. 1, April 1983, pp. 5.1-5.11.

Badger, M. A., "The Assembly Robot—Application in Flexible Assembly Systems," *Proc. 2nd Eur. Conf. Automated Manufacturing*, Birmingham, UK, May 1983, pp. 243-250.

Bailey, J. R., "Product Design for Robotic Assembly," *Proc. 13th Int. Symp. Industrial Robots* and *Robots 7*, Chicago, April 1983, Vol. 1, pp. 11.44-11.57.

Beck, F., et al., "Programmable Assembly with Industrial Robots—Components, Concepts and Solutions," *Proc. 14th Symp. Industrial Robots/7th Int. Conf. Industrial Robot Technology*, Gothenburg, Sweden, Oct. 1984, pp. 461–474.

Bloom, H. J., "Robotic Diskette Writer System," *Proc. 5th Int. Conf. Assembly Automation*, Paris, May 1984, pp. 35–44.

Bobe, U., Optiz, H., and Simon, U., "Use of Robots in Assembly without Human Operators," *Wiss. Z. Tech. Hochsch. Karl-Marx-Stadt*, Vol. 24, No. 5, 1982, pp. 568–572 (in German).

Bortz, A. B., "Robots in Batch Manufacturing. The Westinghouse APAS Project," *Robot. Age*, Vol. 6, No. 1, 1984, pp. 12–18.

Camera, A. and Arban, A., "Examples of Assembly Robot Applications in the Manufacturing Industry," *Proc. 5th Int. Conf. Assembly Automation*, Paris, May 1984, pp. 4554.

Cannon, K. C., "Adapting Robotics and Computer Vision Technologies to Hybrid Assembly," *Proc. 33rd Electronic Components Conf.*, Orlando, Fla., May 1983, pp. 448–463.

Del Gaudio, I. and Lambiase, A., "Automatic Manufacturing Improvement through Quality Control," *Assem. Autom.*, Vol. 2, No. 2, 1982, pp. 98–101.

Dumse, R. M., "A Robotic Application for Contamination Free Assembly," *J. Forth Appl. Res.*, Vol. 1, No. 1, 1983, pp. 33–41.

Duncheon, C. S., "On Their Own, Robots, Build Winchester Disk Drives," *Electronics*, Vol. 56, No. 2, 1983, pp. 131–132.

Earle, D., "Characteristics and Constraints of Robotic Assembly: A Primer Project Engineering/Management," *Proc. Northcon/83 Electronics Show and Convention*, Portland, Ore., May 1983, pp. 8/2/1–11.

Frommhold, J., "Product Design and Tolerances for Optimum Use of Industrial Robot Assembly," *Wiss. Z. Tech. Hochsch. Karl-Marx-Stadt*, Vol. 24, No. 5, 1982, pp. 573–577 (in German).

Gehrmann, W., "IRB 95—A Set of Adapted Constructive Units for Industrial Robots to Rationalize Assembling Processes," *Fernmeldetechnik*, Vol. 23, No. 2, 1983, pp. 74–77.

Hartley, J., "Robot Assembly Machines," *Assem. Autom.*, Vol. 3, No. 4, 1983, pp. 232–236.

Hartley, J., "Robots Start to Assemble Watches," *Assem. Autom.*, Vol. 3, No. 3, 1983, pp. 169–170.

Heginbothma, W. B., "Programmable Assembly," *Int. Trends Manuf. Technol. Ser.*, p. 360.

Hirabayashi, H., "Travelling Assembly Robot," *Proc. 13th Int. Symp. Industrial Robots and Robots 7*, Chicago, April 1983, Vol. 2, pp. 20.19–20.31.

Hoffman, B. D., et al., "Vibratory Insertion Process: A New Approach to Non-Standard Component Insertion," *Robots 8, Proc. 8th Int. Conf. Industrial Robots*, Detroit, June 1984, Vol. 1, pp. 8.1–8.10.

316   Chapter 9

Holmqvist, U., "Programmable Automatic Assembly Station with New
    ASEA Robot," *Proc. 14th Int. Symp. Industrial Robots/7th Int.
    Conf. Industrial Robot Technology*, Gothenburg, Sweden, Oct.
    1984, pp. 617–622.
Kaftan, R. L., Majewski, F. T., and Riggs, C. D., "Back Panel
    Connector Assembly Using a Robot," *IBM Tech. Disclosure Bull.*,
    Vol. 26, No. 6, 1983, pp. 3078–3080.
Kirsch, K. F., "Pick and Place Robots in Small Part Assembly Sys-
    tems," *Robot Appl. Conf.*, Cleveland, June 1983, pp. M583-454/
    1–13.
Kohno, M., Horino, H., and Isobe, M., "Intelligent Assembly
    Robot," Vol. 30, No. 4, 1981, pp. 211–216. (Also in *Robot
    Vision*, Pugh, A. (Ed.), pp. 201–208.
Korsakov, V. S., et al., "Assembly with Threaded Fasteners Using
    Robots," *Mekh. Avtom. Proizvod.*, Vol. 10, 1983, pp. 2–6
    (in Russian).
Krackle, J. L., "Hydraulic Robotic System Automates Assembly of
    Printer Cartridge Ribbons," *Hydraul. Pneumat.*, Vol. 36, No. 3,
    1983, pp. 61–63.
Leete, M. W., "Integration of Robot Operations at Flymo—A Case
    Study," *Proc. 6th Brit. Robot Assoc. Annu. Conf.*, Birming-
    ham, UK, May 1983, pp. 113–124.
Marko, M. P., "Automated Ultrasonic Insertion at the Amprobe In-
    strument Division of Core Industries," *Proc. Robot Appl. Conf.*,
    Cleveland, June 1983, pp. M83–453/1–5.
Martel, J. J., "The Operation of an Automatic System for the 'Quan-
    tity Production' of Made-to-Measure Motors," *Equip. Electr.
    Electron.*, Vol. 497, 1983, pp. 25–28 (in French).
Merlet, J. C., "A New Method for Inserting a Peg in a Hole Using
    a Robot," *Proc. 4th Int. Conf. Assembly Automation*, Tokyo,
    Oct. 1983, pp. 400–402.
Nakamura, A., *Nat. Tech. Rep.*, Vol. 29, No. 4, 1983, pp. 518–
    527.
Nunnally, H. N. and Craze, E. G., "Analytical Balance Operation
    with an Assembly Robot," *Proc. 13th Int. Symp. Industrial
    Robots and Robots 7*, Chicago, April 1983, Vol. 1, pp. 10.31–
    10.41.
Osorio, A., "Workpiece Identification, Grasping and Manipulation in
    Robotics," *Proc. 5th Int. Conf. Assembly Automation*, Paris,
    May 1984, pp. 143–152.
Ramsden, N. C., Shanks, L., and Rathmill, K., "The Development
    of Techniques for the Robot Automated Assembly of Sheet
    Metal," *Proc. 5th Int. Conf. Assembly Automation*, Paris, May
    1983, pp. 191–206.
"A Robotic Wire Harnessing System," *Assem. Autom.*, Vol. 3, No.
    1, 1983, pp. 44–48.

Schmieder, L., "The Assembly of an Oilpump with Cooperation of an Industrial Robot, a Supervising Real-Time Processor and a Tactile Sensor," *Proc. Robots Automotive Industry Int. Conf.*, Birmingham, UK, April 1982, pp. 157–162.

Scott, P. B. and Husband, T. M., "Robotic Assembly: Design, Analysis and Economic Evaluation," *Proc. 13th Int. Symp. Industrial Robots* and *Robots 7*, Chicago, April 1983, Vol. 1, pp. 5.12–5.29.

Smith, R. C. and Nitzan, D., "A Modular Programmable Assembly Station," *Proc. 13th Int. Symp. Industrial Robots* and *Robots 7*, Chicago, April 1983, Vol. 1, pp. 5.53–5.62.

Stamp, P. B., "Automatic Assembly of Pneumatic Valves," *Proc. 7th Brit. Robot Assoc. Annu. Conf.*, Cambridge, UK, May 1984, pp. 193–202.

Stepourjine, R. and Rouget, J. P., "Automatic Insertion Module for Light Robotics," *Dev. Robot.*, 1983, pp. 197–204.

Takahashi, M. and Kohno, M., "An Assembly Robot System with Twin Arms and Vision," *Proc. 12th Int. Symp. Industrial Robots/6th Conf. Industrial Robot Technology*, Paris, June 1982, pp. 111–120.

Takemura, K. and Kubota, K., "Automatic Assembly Systems Using Robots," *Nat. Tech. Rep.*, Vol. 29, No. 4, 1983, pp. 544–551 (in Japanese).

Uno, M., Miyakawa, A., and Ohashi, T., "Multiple Parts Assembly Robot Station with Visual Sensor," *Proc. 5th Int. Conf. Assembly Automation*, Paris, May 1984, pp. 55–64.

VanderBrug, G. J., "Programmable Assembly with Vision," *Assem. Autom.*, 1983, pp. 92–97.

VanderBrug, G. J., Wilt, D., and Davis, J., "Robotic Assembly of Keycaps to Keyboard Arrays," *Proc. 13th Int. Symp. Industrial Robots* and *Robots 7*, Chicago, April 1983, Vol. 1, pp. 5.40–5.52.

Wright, A. J., "Light Assembly Robots—An End Effector Exchange Mechanism," *Mech. Eng.*, Vol. 105, No. 7, 1983, pp. 29–35.

Automobile Industry

Astrop, A., "Cars That Help Build Cars at Fiat," *Mach. Prod. Eng.*, Vol. 140, No. 3602, 1982, pp. 19–21.

Bleuler, U. and Rudin, O., "Flexible Automation in Motor Vehicle Manufacture," *Werkstatt Betr.*, Vol. 116, No. 6, 1983, pp. 337–340 (in German).

Brunk, W., "Geometric Control by Industrial Robots," *Proc. 2nd Int. Conf. Robot Vision Sensory Controls*, Stuttgart, West Germany, Nov. 1982, pp. 223–231.

Cambert, L. I., "Robots as Key-Equipment in a Highly Mechanized Body Plant in the Volvo Car Corporation, Gothenburg," *Proc. 14th Int. Symp. Industrial Robots/7th Int. Conf. Industrial Robot Technology*, Gothenburg, Sweden, Oct. 1984, pp. 167–174.

Christen, W., "Industrial Robot Serves De-Flashing Stations," *Kunst. Ger. Plast.*, Vol. 74, No. 3, 1984, pp. 8–9.

Ericsson, S., "Automated Engine Assembly with Industrial Robots at Saab-Scania," *Proc. 14th Int. Symp. Industrial Robots/7th Int. Conf. Industrial Robot Technology*, Gothenburg, Sweden, Oct. 1984, pp. 251–260.

Garrett, K., "Sniffing Robots Check Cars for Water Leaks," *Meas. Inspect. Technol.*, Vol. 4, No. 8, 1982, pp. 23–24.

Granstedt, O., "ASEA IRb in Automotive Industry," *Proc. 6th Brit. Robot Assoc. Annu. Conf.*, Birmingham, UK, May 1983, pp. 79–90.

Gzik, H., "Car Body Work in Small and Medium Series with Industrial Robots," *VDI Z.*, Vol. 125, No. 13, 1983, pp. 505–515.

Huang, B. and Ruddle, J. L., "Requirements of Sealant Application Robots," *Robots 8, Proc. 8th Int. Conf. Industrial Robots*, Detroit, June 1984, Vol. 1, pp. 7.1–7.7.

Johnson, R. L., "A Robot Solution to Liquid Sealer Application," *Robots 8, Proc. 8th Int. Conf. Industrial Robots*, Detroit, June 1984, Vol. 1, pp. 7.8–7.21.

LeCerf, B. H. and Berry, B. H., "Nissan's Truck Plant: People and Robots Under One Roof," *Iron Age*, Vol. 225, No. 26, 1982, pp. 29–30,33.

Maali, F., Besant, C. B., "Visual Recognition of Engine Bearing Caps," *Proc. 4th Int. Conf. Robot Vision Sensory Controls*, London, Oct. 1984, pp. 427–436.

Meunier, J., "Automation of the Douai Factory: Assembling Renault 9 Automobile Body," *Rev. Gen. Electr.*, Vol. 11, 1982, pp. 753–759 (in French).

Muller, S., "Car-Body Construction and Assembly," (KUKA Schweissanlagen & Roboter, West Germany, *Assem. Autom.*, Vol. 2, No. 2, 1982, pp. 110–113.

Persson, J. C., "Industrial Robots in Competitive Automotive Components Manufacturing," *Proc. 14th Int. Symp. Industrial Robots/7th Int. Conf. Industrial Robot Technology*, Gothenburg, Sweden, Oct. 1984, pp. 175–182.

Rooks, B. W., "Flexibility Can Be Achieved through Robot Assembly," *Assem. Automat.*, Vol. 3, No. 4, 1983, pp. 218–221.

Schupp, G., "Flexible Assembly Automation in the Automobile Industry—Examples of Application and Concepts," *ZWF Z. Wirtsch. Fertigung*, Vol. 78, No. 11, 1983, pp. 493–497 (in German).

Warnecke, H. J. and Walther, J., "Assembly of Car Engines by Hand in Hand Working Robots," *Proc. 4th Int. Conf. Assembly Automation*, Tokyo, Oct. 1983, pp. 15-23.

Warnecke, H. J., Abele, E. and Walther, J., "Programmable Assembly Cell for Automotive Parts and Units," *Proc. '83 Int. Conf. Advanced Robotics*, Sept. 1983, Pt. 1, pp. 29-38.

Warnecke, H. J., et al., "Application of Industrial Robots for Assembly—Operations in the Automotive Industry," *Proc. 13th Int. Symp. Industrial Robots and Robots 7*, Chicago, April 1983, Vol. 1, pp. 5.30-5.39.

Casting and Molding

Baumann, R. D. and Wilmshurst, D. A., "Vision System Sorts Castings at General Motors Canada," *Sensor Rev.*, Vol. 2, No. 3, 1982, pp. 145-149. (Also in Engineers Digest, Vol. 28, No. 6, 1982, and *Robot Vision*, pp. 255-266.)

Booth, D. E., "Robots—Their Use in Metal Stamping," *Sheet Met. Ind.*, 1981, pp. 58(11/12), 1982, p. 59(2).

Ferloni, A., "Towards the Automated Factory, Advanced Robotised Lines in Forging Plants," *Dev. Robot. 1983*, Bedford, UK, 1983, pp. 9-18.

Fink, R., "Unloading Devices for Die Casting Machines," *Giessereipraxis*, Vol. 4, 1982, pp. 58-62 (in German).

"Industrial Robots in Sheet Metal Presswork," *Sheet Met. Ind.*, Vol. 60, No. 4, 1983, pp. 213, 215-216.

James, C. F. and Sylvia, J. G., "Robot's Role in Foundry Mechanization," *Mod. Cast.*, Vol. 72, No. 5, 1982, pp. 30-33.

Khar'kov, V. I., "Automatic Manipulators (Robots)," *Mekh. Avtom. Proizvod.*, Vol. 9, 1983, pp. 29-31.

Maiette, R. L., "Plastic Modling, Inserts, and Flexible Automation," *Proc. 8th Int. Conf. Industrial Robots*, Detroit, June 1984, Vol. 1, pp. 1.1-1.5.

Moreau, J. L., "Robotics and Plastic Injection—from Manipulator to Programmable Robot," *Dev. Robot. 1983*, pp. 3-8.

Quinlan, J. C., "Robotic Upset Forging," *Tool. Prod.*, Vol. 49, No. 2, 1983, pp. 79-80.

"Robotic Press Tending Cuts Costs of Oven Liners," *Tool. Prod.*, Vol. 49, No. 6, 1983, pp. 105-106.

"Shell Moulding at Kohlswa Jernverks AP," *Foundry Trade*, Vol. 15, No. 5, 1981, pp. 721, 723-724.

Walton, H. W., "Robots in Heat Treatment," *Heat Treat. Met.*, Vol. 10, No. 4, 1983, pp. 103-107.

Watts, P. L., "Automated Pressure Diecasting," *Proc. 6th Brit. Robot Assoc. Annu. Conf.*, Birmingham, UK, May 1983, pp. 143-148.

Electronic Industry

De la Cruz, M., "The Design of High Performance Robotic Assembly
    Centres for PCB Non-Standard Electronic Component Assembly,"
    Proc. 5th Int. Conf. Assembly Automation, Paris, May 1984,
    pp. 297–308.
Hartley, J., "Omega—Toshiba's Route to Automation," *Assem.
    Autom.,* Vol. 3, No. 3, 1983, pp. 164–168.
Hastie, W. M., "Robotic Assembly for Electronics," *Circuits Manuf.,*
    Vol. 23, No. 1, 1983, pp. 34–40.
Henderson, J. A. and Hoiser, R. N., "New Robotic Systems Change
    the Electronic Assembly Factory," *Robots 8, Proc. 8th Int.
    Conf. Industrial Robots,* Detroit, June 1984, Vol. 1, pp. 8.57–
    8.75.
Hill, J. J., Burgess, D. C., and Pugh, A., "The Vision-Guided
    Assembly of High Power Semiconductor Diodes," *Proc. 14th Int.
    Symp. Industrial Robots/7th Int. Conf. Industrial Robot Tech-
    nology,* Gothenburg, Sweden, Oct. 1984, pp. 449–460.
Iscoff, R., "Robots on the Line," *Electron. Packag. Prod.,* Vol. 23,
    No. 4, 1983, pp. 49–53 and pp. 80–86.
Jones, S. T., "Turnkey Systems for Progressive Batch PCB Assem-
    bly and Inspection," *Proc. 3rd Int. Conf. Flexible Manufactur-
    ing Systems* and *17th Annu. IPA Conf.,* Boeblingen, West Ger-
    many, Sept. 1984, pp. 431–440.
Kishi, G. T. and Frohlich, J. G., "Robot System for Assembly and
    Test of Printed Circuit Motor Armatures," *Robots 8, Proc. 8th
    Int. Conf. Industrial Robots,* Detroit, June 1984, Vol. 1, pp.
    10.16–10.30.
Kocherscheidt, G. and Urban, S., "Portal Robots with Step Motor
    Drive for Assembly and Printed Board Drilling," *Feinwerktech.
    Messtech.,* Vol. 91, No. 6, 1983, pp. 266–268 (in German).
Kohno, M., et al. "Application of a Robot to Adjustment and Test-
    ing of Electronic Goods," *Proc. 14th Int. Symp. Industrial
    Robots/7th Int. Conf. Industrial Robot Technology,* Gothenburg,
    Sweden, Oct. 1984, pp. 541–552.
Lyman, J., Keller, E. L., and Comerford, R., "Manufacturing Tech-
    nology in Electronics," *Electronics,* Vol. 56, No. 20, 1983, pp.
    164–187.
Mangin, C. H. and D'Agostino, S., "Flexible Assembly for the
    Electronic Industry," *Assem. Autom.,* Vol. 3, No. 3, 1983, pp.
    139–142.
McCleary, R. D., "Printed Circuit Board Assembly and Test: Is
    Automation Adaptable?," *Robots 8, Proc. 8th Int. Conf. Indus-
    trial Robots,* Detroit, June 1984, Vol. 1, pp. 8.76–8.85.
Murphy, A. D. and Kenney, A. G., "Using Robots to Optimise
    Automated PCB Assembly," *Proc. 7th Brit. Robot Assoc. Annu.
    Conf.,* Cambridge, UK, May 1984, pp. 183–192.

Riboulet, P., "A Robot Application at the IBM France, Bordeaux, Plant and Its Economic Aspects," Productivity and Data Processing: Two Essentials for a Dynamic Company, *Proc. Spring. Convention*, Paris, May 1983, Vol. 1, pp. 104–106.

Smith, B. S., "Industrial Robot Speeds PWB Fabrication," *Electronics*, Vol. 29, No. 8, 1983, pp. 53–55.

Sweeney, E., Redmon, P., and Harhen, J., "FMS in Electronic Assembly—A Case Study," *Proc. 3rd Int. Conf. Flexible Manufacturing Systems* and *17th Annu. IPA Conf.*, Boeblinger, West Germany, Sept. 1984, pp. 467–476.

Van Brussel, H. and Thielemans, H., "Laser Guided Robot Assembly of Printed Circuit Boards," *Proc. 5th Int. Conf. Assembly Automation*, Paris, May 1984, pp. 75–84.

Finishing and Painting

Bartlam, P. and Neilson, G., "Vision System to Identify Car Body Types for a Spray Painting Robot," *Proc. 3rd Int. Conf. Robot Vision Sensory Controls*, Cambridge, Mass., Nov. 1983, pp. 517–522.

Dooley, R. A., "Cartesian Paint Spraying Robots—Installation Arrangement Possibilities," *Proc. 8th Int. Conf. Industrial Robots*, Detroit, June 1984, Vol. 1, pp. 7.22–7.31.

Dust, A. G., "Integration of Robot in Solution Dip Operations," *Proc. 13th Int. Symp. Industrial Robots and Robots 7*, Chicago, April 1983, Vol. 1, pp. 2.32–2.40.

Ferretti, M., "Laborious, Dangerous and Noxious, Surface Treatments are Using Robots," *Nouv. Autom.*, Vol. 28, No. 38, 1983, pp. 39–43 (in French).

"Finishing in the Automotive Industry," *Prod. Finish.*, Vol. 35, No. 11, 1982, pp. 20, 22, 38.

Fischer, P. and Schmidt, J., "Application of Industrial Robots for Deburring of Castings," *Giessereitechnik*, Vol. 28, No. 11, 1982, pp. 339–441 (in German).

Fredriksson, L. B., "Robotics in a Spray-up Process," *Proc. 14th Int. Symp. Industrial Robots/7th Int. Conf. Industrial Robot Technology*, Gothenburg, Sweden, Oct. 1984, pp. 297–304.

Goodwin, J. G., "Finishing Robots—Towards 100% Up Time", *Proc. 13th Int. Symp. Industrial Robots and Robots 7*, Chicago, April 1983, Vol. 1, pp. 8.12–8.23.

Graham, D., Jenkins, S. A., and Woodwark, J. R., "Model Driven Vision to Control a Surface Finishing Robot," *Proc. 3rd Int. Conf. Robot Vision Sensory Controls,"* Cambridge, Mass., Nov. 1983, pp. 433–440.

Grunewald, P., "Car Body Painting with the Spine Spray System," *Proc. 14th Int. Symp. Industrial Robots/7th Int. Conf. Industrial Robot Technology*, Gothenburg, Sweden, Oct. 1984, pp. 633–642.

Gustafsson, L., "Cleaning of Castings—a Typical Job for a Robot," *Proc. 13th Int. Symp. Industrial Robots and Robots 7*, Chicago, April 1983, Vol. 1, pp. 8.24–8.33.

Gustafsson, L., "Cleaning of Castings—a Typical Job for a Robot," Society of Manufacturing Engineers Tech. Paper Ser. MS 83–347, 1983.

Irons, C. and George, L., "Thermal Spray Robots," *Proc. 13th Int. Symp. Industrial Robots and Robots 7*, Chicago, April 1983, Vol. 1, pp. 8.34–8.51.

Kelly, M. P., "Robots in the Paint Shop," *Proc. 6th Brit. Robot Assoc. Annu. Conf.*, Birmingham, UK, May 1983, pp. 71–77.

"Mitsubishi-Iwata Paint Coating Robot," *Tech. Rev. Mitsubishi Heavy Ind.*, Vol. 19, No. 3, 1982, pp. 269–271.

Mortensen, A., "Automatic Grinding," *Proc. 13th Int. Symp. Industrial Robots and Robots 7*, Chicago, April 1983, Vol. 1, pp. 8.1–8.11.

"New Robotic NC System Paints GM Auto Bodies," *Tool. Prod.*, Vol. 49, No. 6, 1983, pp. 87–88, 90, 92.

"Paint Spraying," *Finishing*, Vol. 6, No. 12, 1982, pp. 14–16.

Polak, M., "Technological Prerequisites for the Application of Industrial Robots for Grinding of Castings," *Giessereitechnik*, Vol. 28, No. 11, 1982, pp. 334–336 (in German).

Robert, N., et al., "Adaptive Burring Tool for a Robot Finishing Plastic Pieces," *Proc. 7th Brit. Robot Assoc. Annu. Conf.*, Cambridge, UK, May 1983, pp. 203–214.

Rutledge, R., "Interfacing the Paint Robot into the Manufacturing Process," *Robot Appl. Conf.*, Cleveland, June 1983, pp. M583–455/10.

Schraft, R. D., et al., "Application of Sensor Controlled Robots for Fettling of Castings," *Proc. 13th Int. Symp. Industrial Robots and Robots 7*, Chicago, April 1983, Vol. 2, pp. 13.44–13.57.

Spur, G. and Felsing, W., "Development of a Robot-Integrated Working–Place for the Brushing of Rubber Metal Parts," *Proc. 13th Int. Symp. Industrial Robots and Robots 7*, Chicago, April 1983, Vol. 1, pp. 8.52–8.60.

"Swedish Foundry Installs Robot for Cleaning Castings," *Foundry Trade*, Vol. 154, No. 3266, 1983, pp. 841–842.

Sylvia, J. G. and James, C. F., "Robot/Plasma-Arc Interface for Cleaning Gray Iron Castings," *Proc. 6th Brit. Robot Assoc. Annu. Conf.*, Birmingham, UK, May 9183, pp. 167–173.

Thenander, R., "Practical Examples of Deburring with ASEA Robot," *Proc. 6th Brit. Robot Assoc. Annu. Conf.*, Birmingham, UK, May 1983, pp. 101–111.

Williams, D. J. and Phillips, R. G., "Robotic Deburring of Connect-
ing Rod Linear Slots," *Proc. 7th Brit. Robot Assoc. Annu.
Conf.*, Cambridge, UK, May 1984, pp. 215–222.
Woolley, E., "Robots, Automation and Spray Painting," *Ind. Manage.
Data Syst.*, Sept. 1983, pp. 7–9.

Hazardous Environments

Bohm, H. and Lauer, P., "Tele-Operated Manipulator for the Test-
ing and Repair of Steam Generator Heating Pipes," *Jahrestagung
Kerntechnik '84 (Annu. Meet. Nucl. Technol.)*, Frankfurt, May
1984, pp. 599–602.
Comeau, J. L., "Clean-Room Robot Implementation," General Electric
Report No. GEPP-TIS-663, 1982.
Duggan, F., et al., "A Nuclear Application: Hardware Simulation
of Decontamination by a Robot," *Proc. 7th Brit. Robot Assoc.
Annu. Conf.*, Cambridge, UK, May 1984, pp. 75–84.
Frandscen, G. B., "Electromechanical Manipulator for HCVF," *Trans.
Am. Nucl. Soc. 1982 Winter Meet.*, Washington, D.C., Nov.
1982, Vol. 43, pp. 754–755.
Hasegawa, T., et al., "A Survey on the Application of Robot Tech-
niques to an Atomic Power Plant," *Bull. Electrotech. Lab.*, Vol.
46, No. 9, 1982, pp. 468–487.
Hamel, W. R. and Martin, H. L., "Robotics-Related Technology in
the Nuclear Industry," Department of Energy Report CONF-
830874-22, 1983.
Holt, R. C., "Evaluation of Angle Transducers for a High Perform-
ance Manipulator," *Rec. Colloq. Axes Measurement Systems for
Robots, Manipulators and Other NC Devices*, London, March
1983, p. 3.
Kashihara, H., et al., "Servomanipulator Having New Multi-Joint
and Ingenious Wrist," *Proc. '83 Int. Conf. Advanced Robotics*,
Sept. 1983, Pt. 1, pp. 59–66.
Kemmochi, S. and Kazuoka, S., "Mechanized Devices for the Inserv-
ice Inspection of Nuclear Power Plants," *Proc. '83 Int. Conf.
Advanced Robotics*, Sept. 1983, Pt. 1, pp. 83–90.
Leary, R. H., et al., "Remote Demilitarisation of Chemical Muni-
tions," *Robots 8, Proc. 8th Int. Conf. Industrial Robots*, De-
troit, June 1984, Vol. 1, pp. 9.28–9.41.
Lowe, D. B., "Robot Technology in Remote Inspection and Repair,"
Meas. Control, Vol. 14, No. 9, 1981, pp. 283–287.
Martin, H. L. and Hamel, W. R., "Joining Teleoperation with Ro-
botics for Remote Manipulation in Hostile Environments," *Robots
8, Proc. 8th Int. Conf. Industrial Robots*, Detroit, June 1984,
Vol. 1, pp. 9.1–9.17.

Silverman, E. B., et al., "Industrial Remote Inspection System (IRIS)," *Robots 8, Proc. 8th Int. Conf. Industrial Robots,* Detroit, June 1984, Vol. 1, pp. 9.42–9.48.

Stahn, H., "Microprocessor Control of Industrial Robots," *Wiss. Z. Tech. Univ. Dresden,* Vol. 30, No. 213, 1981, pp. 231–238 (in German).

Tomizawa, F., Ozaki, N., and Sato, C., "Infrared Wireless Telecommunication System for a Remotely Controlled Robot in Nuclear Power Plant," *Trans. Am. Nucl. Soc. 1983 Annu. Meet.,* Detroit, June 1983, Vol. 44, pp. 576–577.

Yount, J. A., "Microprocessor Enhanced Real Time Manual Control of an Industrial Robot," *Robots 8, Proc. 8th Int. Conf. Industrial Robots,* Detroit, June 1984, Vol. 1, pp. 9.18–9.27.

Inspection and Testing

Braggins, D., "An Eye for Inspection," *Eng. Dig.,* Vol. 44, No. 5, 1983, pp. 31–34.

Chen, M. J., "Vision Guided X-Y Table for Inspection," *Proc. SPIE Int. Soc. Opt. Eng. Conf. Robotics Industrial Inspection,* San Diego, Aug. 1982, Vol. 360, pp. 287–296.

Curtis, S. B., "Computer Vision Inspection from a CAD Database," *Proc. 8th Int. Conf. Industrial Robots,* Detroit, June 1984, Vol. 1, pp. 10.45–10.54.

Ferretti, M., "From CAOA up to Measurement Robots: Quality First," *Nouv. Autom.,* Vol. 28, No. 37, 1983, pp. 37–44.

Heywood, P. W. and Upcott, D. N., "The Application of Automatic Visual Inspection Systems," *Proc. 7th Brit. Robot Assoc. Annu. Conf.,* Cambridge, UK, May 1984, pp. 295–308.

Huang, G., "Robotic Alignment and Inspection System for Semi-Conductor Processing," *Proc. 3rd Int. Conf. Robot Vision Sensory Controls,* Cambridge, Mass., Nov. 1983, pp. 563–572.

Langdon, R., "Robots Have a Nose for Quality," *Metalwork. Prod.,* Vol. 129, No. 9, 1982, pp. 90–91.

Lowe, H., "An Integrated Robot Vision and Tactile Automotive Inspection System," *Proc. Int. Conf. Advances in Manufacturing,* Singapore, Oct. 1984, pp. 247–258.

Mazur, H. J., Muller, C., and Muller, K. D., "Industrial Robot in Automatic Testing," *Elektronik,* Vol. 32, No. 7, pp. 69–73 (in German).

Schmidberger, E. J. and Ahlers, R. J., "Quality Control with a Robot-Guided Electro-Optical Sensor," *Proc. 4th Int. Conf. Robot Vision Sensory Controls,* London, Oct. 1984, pp. 27–36.

Swenson, L., "How to Add Smart Stuff to a Dumb Robot or a Simple Fail–Soft Robot Application for a Visual Inspection Station," *Proc. 8th Int. Conf. Industrial Robots,* Detroit, June 1983, Vol. 1, pp. 3.47–3.57.

"Test Piece Bad, Test Piece Good. Robot Tests Printed Circuit Boards," *Elektronik*, Vol. 32, No. 16, 1983, pp. 79–81 (in German).

Thibadeau, R., "Printed Circuit Board Inspection," *Carnegie-Mellon Univ. Rep. CMU-RI-TR-81-8*, 1981.

Thibadeau, R., Friedman, M., and Seto, J., "Automatic Inspection for Printed Wiring," *Carnegie-Mellon Univ. Rep. CMU-RI-TR-82-16*, 1982.

Trombly, J. E., "A New Approach to Machine Vision Simplifies Application Development," *Electro '83 Electronics Show and Convention Proc.*, New York, April 1983, pp. 20/5/1–8.

Wickert, L., "Ultrasonic Measuring Tests Using an Industrial Robot," *Ind.-Anz.*, Vol. 105, No. 78, 1983, pp. 38–39.

Machining

Abele, E., Boley, D., and Sturz, W., "Interactive Programming of Industrial Robots for Deburring," *Proc. 14th Int. Symp. Industrial Robots/7th Int. Conf. Industrial Robot Technology*, Gothenburg, Sweden, Oct. 1984, pp. 505–516.

Arnold, K. and Edwards, R., "Robotic Applications at Texas Instruments, Sherman Facility," *Proc. 13th Int. Symp. Industrial Robots and Robots 7*, Chicago, April 1983, Vol. 0, pp. 2.1–2.9.

"Automated Machining of Turbine Blades by Rolls-Royce," *Aircr. Eng.*, Vol. 55, No. 1, 1983, pp. 7–9.

Cartwright, R., "Robotic Tool Changing," *Proc. 2nd Eur. Conf. Automated Manufacturing*, Birmingham, UK, May 1983, pp. 219–222.

Chelidze, G., Tordia, P., and Dolidze, T., "Automatized Robotized Line for Machining of Cylindrical Units," *Proc. 6th Brit. Robot Assoc. Annu. Conf.*, Birmingham, UK, May 1983, pp. 137–144.

Horng, S. Y., "Optimisation of Robotic Drilling System," *Proc. 8th Int. Conf. Industrial Robots*, Detroit, June 1984, Vol. 1, pp. 17.9–17.21.

Landsell, G., "Flexible Machining Cell for Turning with a Robot for Change of Set-up, Parthandling and Measuring," *Proc. 14th Int. Symp. Industrial Robots/7th Int. Conf. Industrial Robot Technology*, Gothenburg, Sweden, Oct. 1984, pp. 305–314.

Leckie, R. S., "Parametric Design of Robotic Tooling," *Proc. 13th Int. Symp. Industrial Robots and Robots 7*, Chicago, April 1983, Vol. 1, pp. 7.16–7.34.

Logan, J. C., "Medium Technology Robots for Dome Cylinder Machining System," *Proc. 8th Int. Conf. Industrial Robots*, Detroit, June 1984, Vol. 1, pp. 1.26–1.44.

Molander, T., "Routing and Drilling with an Industrial Robot," *Proc. 13th Int. Symp. Industrial Robots and Robots 7*, Chicago, April 1983, Vol. 1, pp. 3.37–3.47.

Mortensen, A., "Automatic Grinding," Society of Manufacturing Engineers Tech. Paper Ser. MS 83-345, 1983.

Ostby, K., "Robotic Water Jet Cutting," *Proc. 8th Int. Conf. Industrial Robots*, Detroit, June 1984, Vol. 1, pp. 5.26-5.38.

Romberg, D. and Unger, K., "Automatic Tube Machining—The Contribution of Robot Technology in Chemical Plant Construction," *Fertigungstech. Betr.*, Vol. 33, No. 7, 1983, pp. 414-416 (in German).

Swarup, S., "Robot Slashes Machining Cost," *Robot. Today*, Vol. 5, No. 2, 1983, pp. 64-65.

Wolf, H. J., Hausler, W., and Schreiner, J., "Experiences in Technology in Utilization of Industrial Robots with CNC Turning Machines," *Fertigungstech. Betr.*, Vol. 33, No. 10, 1983, pp. 591-594 (in German).

Materials Handling

Abair, D. W., "Modern Solutions to Old Problems—Palletizing with Industrial Robots," *Robots 8, Proc. 8th Int. Conf. Industrial Robots*, Detroit, June 1984, Vol. 1, pp. 3.29-3.46.

Aldridge, M. E., "Robot Unloading of Injection Moulding Machines—Part of the Survival Package," *Proc. 6th Brit. Robot Assoc. Annu. Conf.*, Birmingham, UK, May 1983, pp. 149-158.

Becher, H., et al. "New Handling Systems as Technical Support for the Working Process. Part 6: Feeding Devices," NASA Report BMFT-FB-HA-82-002, 1980 (in German).

Corrado, J. K., "Material-Handling Robot Works for a Dollar an Hour," *Des. News*, Vol. 39, No. 16, 1983, pp. 120-122.

Drexel, P., "Application Aspects of Modular Manipulator Systems," *Werkstatt Bert.*, Vol. 115, No. 6, 1982, pp. 377-381 (in German).

"Enter the Robots", *Cargo Syst. Int.*, Vol. 9, No. 10, 1982, pp. 22-23.

Ford, E., "Component Pick-up and Feed Using 'Seeing Robots'," *Technica*, Vol. 32, No. 9, 1983, pp. 71-73 (in German).

Gandy, T. G., "A Simple Robot System for Loading/Unloading Internal Grinders," *Proc. 13th Int. Symp. Industrial Robots and Robots 7*, Chicago, April 1983, Vol. 1, pp. 2.25-2.31.

Gelbgras, D., et al., "Grabbing Moving Objects with Visual Feedback," *Sensor Rev.*, Vol. 3, No. 2, 1983, pp. 76-80.

Goldschlager, L. M., "Automated Packing by Robot," *Proc. Conf. Computers Engineering 1983*, Sydney, Aust., August–September 2, 1983, pp. 132-136.

Handel, W. and Laux, W., "Design of the Component Flow Interface between Technological Units with Industrial Robots on the One Hand and Transportation, Storage and Transhipment Processes on the Other," *Fertigungstech. Betr.*, Vol. 33, No. 8, 1983, pp. 446-464 (in German).

Hansen, E., "Industrial Robots Automate Flow of Materials in Manufacturing," *ZWFZ Wirtsch. Fertigung*, Vol. 77, No. 7, 1982,
pp. 305–308 (in German).

Kelley, R. B., et al., "Robot System Which Acquires Cylindrical
Workpieces from Bins," *IEEE Trans. Syst. Manage. Cybern.*,
Vol. SMC-12, No. 2, 1982, pp. 204–213. (Also in *Proc. 9th
Int. Symp. Industrial Robots*, Washington, D. C., March 1979,
pp. 339–355.)

Kritzer, E., "One Dozen Enchained Robots: Application of a Chain
of Robots in a Chocolate Packing Plant," *Elektronik*, Vol. 32,
No. 16, 1983, pp. 92–94 (in German).

Muller, W., Terwissen, B., and Bette, B., "Flexible Handling
Equipment for the Automation Measuring of Workpieces," *VDI Z.*,
Vol. 125, No. 7, 1983, pp. 227–231 (in German).

Parker, J. K., Dubey, R., and Paul, F. W., "Robotic Fabric Handling for Automating Garment Manufacturing," *Trans. ASME,
J. Eng. Ind.*, Vol. 105, No. 1, 1983, pp. 21–26.

Rogers, M. A. M., "Five Years Commitment to Robotics," *Proc. 6th
Brit. Robot Assoc. Annu. Conf.*, Birmingham, UK, May 1983,
pp. 7–23.

Saraga, P., et al., "Unpacking and Mounting TV Deflection Units
Using Visually Controlled Robots," *Proc. 3rd Int. Conf. Robot
Vision Sensory Controls*, Cambridge, Mass., Nov. 1983, pp.
541–548.

Schaefer, J. O., "Discussion and Analysis of Conventional Part
Processing vs. a Single Arm Robot-Fed System," *Proc. 4th Int.
Conf. Assembly Automation*, Tokyo, Oct. 1983, pp. 325–332.

Schivarov, N., et al., "Visual System Helps the Robot to Load and
Unload Industrial Workpieces," *Proc. 1st Int. Conf. Automated
Materials Handling*, London, April 1983, pp. 189–198.

Schiwarov, N. S. and Yanakiev, K. I., "Mechanical Handling System for Automatic Quality Grading and Robotised Wall Tile Palletization," *Proc. 5th Int. Conf. Automation in Warehousing*,
Atlanta, Ga., Dec. 1983, pp. 191–198.

Schwind, G., "Robots: A Flexible Solution for Tough Unitizing
Jobs," *Mater. Handl. Eng.*, Vol. 38, No. 9, 1983, pp. 36–40.

Seidel, L. E., "Improving Weave Room Materials Handling," *Text.
Ind.*, Vol. 145, No. 7, 1981, pp. 48–51.

Shron, V. Z., et al., "Industrial Robots as Means of Mechanizing
Refractories Production," *Refractories*, Vol. 22, Nos. 11–12,
1981, pp. 600–603.

Ward, M. R., Rheaume, D. P., and Holland, S. W, "Production
Plant CONSIGHT Installations," *Proc. SPIE Int. Soc. Opt. Eng.
Robotics Industrial Inspection*, San Diego, Aug. 1982, Vol. 360,
pp. 297–305.

Seam Tracking

Budell, R. A., "Robotic MIG Welding Using Vision Seam Tracking,"
  Detroit, June 1984, Vol. 1, pp. 10.31–10.44.
de Keizer, A. and de Grort, R. J., "Laser-Based Arc Welding Sen-
  sor Monitors Weld Preparation Profile," *Sensor Rev.*, Vol. 4, No.
  1, 1984, pp. 8–10.
Karastoyanov, D. N., "Tracking Rectilinear Seams between Planar
  Details in Welding Robots," *Probl. Tekh. Kibern. Robot.*, Vol.
  17, 1983, pp. 54–59 (in Bulgarian).
"Leading the Way on the Robot Weld Guidance," *Industrial Robot*,
  Vol. 10, No. 2, 1983, pp. 104–107.
Masaki, I., et al., "Arc Welding Robot Visually Tracks Its Way
  along Seam," *Weld. Met. Fabr.*, Vol. 50, No. 1, 1982, pp. 17,
  19.
Masaki, I., et al., "Vision Guided Robot System for Arc Welding,"
  Robot Vision, International Trends in Manufacturing Series,
  pp. 179–185.
Morgan, C. G., et al., "Visual Guidance Techniques for Robot Arc-
  Welding," *Proc. 3rd Int. Conf. Robot Vision Sensory Controls*,
  Cambridge, Mass., Nov. 1983, pp. 615–624.
Sawano, S., et al., "A Sealing Robot System with Visual Seam
  Tracking," *Proc. '83 Int. Conf. Advanced Robotics*, Sept. 1983,
  Pt. 1, pp. 351–358.
Smati, Z., et al., "Laser Guidance System for Robots," *Proc. 4th
  Int. Conf. Robot Vision Sensory Controls*," London, Oct. 1984,
  pp. 91–101.
Smati, Z., Smith, C. J., and Yapp, D., "An Industrial Robot Using
  Sensory Feedback for an Automatic Multipass Welding System,"
  Proc. 6th Brit. Robot Assoc. Annu. Conf., Birmingham, UK,
  May 1983, pp. 91–100.
Stauffer, R. N., "Update on Noncontract Seam Tracking Systems,"
  Robot. Today, Vol. 5, No. 4, 1983, pp. 29–34.
Wadsworth, P. K., "Seam Tracking Emerges in US Welding Develop-
  ments," *Industrial Robot*, Vol. 10, No. 2, 1983, pp. 110–113.
Yee Meng Yew, "Laser and Vision Systems in Robotic Applications,"
  Proc. Int. Conf. Advances Manufacturing, Singapore, Oct.
  1984, pp. 213–222.

Welding and Fabrication

Angelov, A. A., et al., "Application of Industrial Robots for Arc
  Welding or Workpieces in Electric Truck Lifting Devices," *Probl.
  Tekh. Kibern. Robot.*, Vol. 17, 1983, pp. 49–53.
Behnisch, H., "Industrial Robots for Welding and Cutting," *Werk-
  statt Betr.*, Vol. 116, No. 7, 1983, pp. 409–412 (in German).

Brosilow, R., "Robot Arc Welders Earn Their Keep," *Weld. Des. Fabr.*, Vol. 55, No. 3, 1982, pp. 67–72.

Clockspin, W. F., et al., "Visually Guided Robot Arc-Welding of Thin Sheet Steel Pressings," *Proc. 12th Int. Symp. Industrial Robots/6th Int. Conf. Industrial Robot Technology*, Paris, June 1982, pp. 225–230.

Fujita, Y., "Welding Robots Make Progress in the Shipyard," *Weld. Met. Fabr.*, Sept. 1983, pp. 377–382.

Gremm, F., "Robot Welding Line with Adapted Conveyor," *Foerdern Heben*, Vol. 33, Nos. 7–8, 1983, pp. 519–521 (in German).

Gzik, H., "Procedure to Investigate the Task Time when Plane Welding with Industrial Robots," *Ind.-Anz.*, Vol. 105, No. 95, 1983, pp. 30–31 (in German).

Gzik, H., et al., "A Fully-Automatic Industrial Robot Workplace for Welding and Grinding the Welds on Sheet-Metal Components," *VDI Z.*, Vol. 125, No. 7, 1983, pp. 223–225 (in German).

Hage, P. M. and Hewit, J. R., "The Robot Welding of Boiler Tube Platens Assisted by Vision Sensing," *Proc. 3rd Int. Conf. Robot Vision and Sensory Controls*, Cambridge, Mass., Nov. 1983, pp. 635–640.

Hage, P. M. and Hewit, J. R., "Visual Feedback in Boiler Tube-Bank Spacer-Welding," *Proc. 2nd Int. Conf. Robot Vision and Sensory Controls*, Stuttgart, West Germany, Nov. 1983, pp. 73–82.

Hartley, J., "BMW's Alternative Robot Technology," *Industrial Robot*, Vol. 10, No. 2, 1983, pp. 114–117.

Hewit, J. R. and Love, J. G., "The Application of Robotic Welding Technology to Shipbuilding," *Proc. 13th Int. Symp. Industrial Robots and Robots 7*, Chicago, April 1983, Vol. 1, pp. 6.70–6.88.

Hoh, T., et al., "The Tack Welding and Assembling Robot," *Proc. '83 Int. Conf. Advanced Robotics*, 1983, Pt. 1, pp. 381–388.

Hughes, R. V., et al., "Robot Plasma Welding with Integral Arc Guidance," *Proc. 4th Int. Conf. Robot Visionand Sensory Controls*, London, Oct. 1984, pp. 123–136.

"Hyster Finds that Robots are Easy but Welding is Hard," *Industrial Robot*, Vol. 10, No. 2, 1983, pp. 100–103.

Kallevig, J. A., "Robotic Welding of Honeywell," *Proc. 13th Int. Symp. Industrial Robots and Robots 7*, Chicago, April 1983, Vol. 2, pp. 6.15–6.25.

Kaufmann, H., "ASEA's Secondary Current Transmission System for Spotwelding Robots," *Dev. Robot.*, 1983, pp. 73–78.

Kerth, W. J. and Kerth, R. J., "Mobile and Stationary Adaptive Welding Systems," *Proc. 8th Int. Conf. Industrial Robots*, Detroit, June 1984, Vol. 1, pp. 11.26–22.40.

330   Chapter 9

Klie, H., "Practical Application of Robotic Arc Welding and Cutting,"
     Proc. 7th Brit. Robot Assoc. Annu. Conf., Cambridge, UK,
     May 1984, pp. 39–48.
Kossin. C., "A Spot Welding Robot with Integrated Secondary Cur-
     rent Conductor System," *Ino.-Anz.*, Vol. 105, No. 7, 1983,
     pp. 46–47 (in German).
LaLoe, D. and Seibert, L., "3D Vision Guided Welding Robot Sys-
     tem," *Industrial Robot*, Vol. 11, No. 1, 1984, pp. 18–20.
Lange, M. S., "Arc Welding with Robots," *Mod. Mach. Shop*, Vol.
     56, No. 6, 1983, pp. 66–76.
Lebedev, V. K., Zaruba, I. I., and Pentegov, I. V., "Development
     Trends in Power Sources for Arc Welding," *Autom. Weld.*, Vol.
     35, No. 8, 1982, pp. 1–7.
Marcotte, D. E., "Robotic Arc Welding in Production," *Robot Appl.*
     Conf., Cleveland, June 1983.
Miller, K. J., "Robot Arc Welding at Caterpillar Tractor Company,"
     Developments in Mechanised, Automated and Robotic Welding,
     Proc. Int. Conf., London, Nov. 1980, pp. 39.1–9.
Monckton, P. S., Hawthorn, R. W., and Hutton, P. M., "A Robot
     Welding Cell for Small Batch Production," *Proc. Int. Conf. Ad-*
     vances Manufacturing, Singapore, Oct. 1984, pp. 223–230.
Monckton, P. S., Hawthorn, R. W., and Jones, R., "Simulation of
     a Robotic Welding Cell to Be Used in Small Batch Production,"
     Proc. 7th Brit. Robot Assoc. Annu. Conf., Cambridge, UK,
     May 1984, pp. 61–74.
Moussalli, P., et al., "Make Simple Automation: Design of a 2 Axis
     Robot for Welding," *Nouv. Autom.*, Vol. 28, No. 42, 1983, pp.
     63–66 (in French).
"Multi-Articulated Type Robot for Arc Welding," *Ishikawajima-Harima*
     Eng. Rev., Vol. 23, No. 2, 1983, pp. 148–154 (in Japanese).
Nishimura, H., et al., "Applications of Cartesian-Type Unit Robots,"
     Nat. Tech. Rep., Vol. 29, No. 4, 1983, pp. 537–543.
"Now It's Robotics for TIG," *Weld. Mat. Fabr.*, June 1983, p. 227.
Pavone, V. J., "User Friendly Welding Robots," *Proc. 13th Int.*
     Symp. Industrial Robots and *Robots 7*, Chicago, April 1983,
     Vol. 1, pp. 6.89–6.102.
Richardson, R. W., et al., "The Feasibility of Robotic Arc Welding
     of a Low Volume Sheet Metal Fabrication," *Proc. 13th Int. Symp.*
     Industrial Robots and *Robots 7*, Ch2cago, April 1983, Vol. 1,
     pp. 6.103–6.112.
Rider, G., "Control of Weldpool Size and Position and Automatic and
     Robotic Welding," *Proc. 3rd Int. Conf. Robot Vision and Sen-*
     sory Controls, Cambridge, Mass., Nov. 1983, pp. 615–634.
"Robotic Welding," *Weld. J.*, Vol. 61, No. 9, 1982.
Sakakibara, S., "The Latest Functions of Spot Welding Robot,"
     Sept. 1983, Pt. 1, pp. 375–380.

Schraft, R. D. and Gzik, H., "Welding with Industrial Robots,"
Schweisstech. Soudure, Vol. 72, No. 9, 1982, pp. 279–286
(in German).

Schraft, R. D., et al., "Flexible Assembly and Flexible—New Developments," *Proc. 3rd Int. Conf. Flexible Manufacturing Systems
and 17th Annu. IPA Conf.*, Boeblingen, West Germany, Sept.
1984, pp. 441–454.

Schivarov, N. and Janakiev, K., "Manipulating System with Automatic Electric Arc Welding of Workpiece by Means of an Industrial Robot," *Proc. 2nd Eur. Conf. Automated Manufacturing*,
Birmingham, UK, May 1983, pp. 443–452.

Smart, R. S., "Flexibility in Resistance Spot Welding," *Proc. 2nd
Eur. Conf. Automated Manufacturing*, Birmingham, UK, May
1983, pp. 117–126.

Sthen, T. and Porsander, T., "An Adaptive Torch Positioner System for Welding of Car Bodies," *Proc. 3rd Int. Conf. Robot
Vision and Sensory Controls*, Cambridge, Mass., Nov. 1983,
pp. 607–614.

Sweet, L. M. and Case, A. W., "Advanced Arc Welding Robot Technology Using Vision Sensor-Based Control Systems," *Proc. 8th
Int. Conf. Industrial Robots*, Detroit, June 1984, Vol. 1, pp.
11.1–11.13.

Tembel, G., "Arc Welding Robots and Their Practical Applications,"
Proc. Int. Conf. Advances Manufacturing, Singapore, Oct.
1984, pp. 231–238.

Thorne, R. G. and Middle, J. E., "The Factors Effecting the Successful Implementation of a Robotic Arc Welding System in Batch
Manufacturing," *Dev. Robot.*, 1983, pp. 19–32.

Villers, P., "Present Industrial Use of Vision Sensors for Robot
Guidance," *Proc. 12th Int. Symp. Industrial Robots/6th Int.
Conf. Industrial Robot Technology*, Paris, June 1982, pp.
291–302.

Wadsworth, P. K., "Robotic Arc Welding in Medium Batch Fabrication
Companies," *Proc. 6th Brit. Robot Assoc. Annu. Conf.*, Birmingham, UK, May 1983, pp. 205–216.

Westberg, C., "Design for Mechanized Welding," *Inst. Verkstadstek.
Forsk. IVF Result. 82615*, June 1982 (in Swedish).

10

Geometry of the Workspace

10.1   ACCESSIBLE AND DEXTROUS WORKSPACE

In order to study the workspace of a robot, the structure of the robot can be considered as consisting of the arm and the hand. The arm is the large regional structure for global positioning of the hand, which is the small orientational structure for orientating the tool. The primary workspace of such a robot with a large regional structure and a small orientational structure is determined by the arm. The hand generates the secondary workspace of a robot.

The workspace of a robot is an important criterion in evaluating manipulator geometries. Manipulator workspaces may be described in terms of the accessible workspace and the dextrous workspace. Accessible workspace is the volume within which every point can be reached by a reference point on the hand. Dextrous workspace denotes an accessible workspace within which the hand can be rotated in any specified orientation. In the dextrous workspace the robot has complete manipulative capability. However, in the accessible workspace, the manipulator's operational capacity is limited because the terminal device can only be placed in a restricted range of orientations.

Figure 10.1 shows the work area of two- and three-link robots. These plots were obtained on a microcomputer programmed so that the numeric keypad could be used to move each link. The procedure for obtaining these work areas is to move each link through its angular range of motion in a manner similar to the physical motion of the robot arm.

The work area of linkage robots driven by hydraulic cylinders depends on how the cylinders are configured to drive the joints.

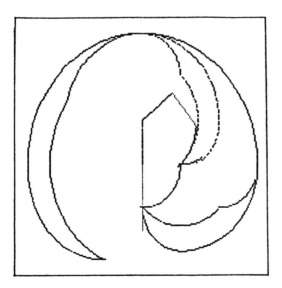

FIGURE 10.1   Developing the workspace of two- and three-link robots on a computer.

Figure 10.2 shows the work area for different configurations developed on an interactive computer graphics system. A design chart for the work area of a two-link robot developed on a computer is shown in Figure 10.3 for different angular ranges of the two joint angles. The work areas of two different robots are shown in Figure 10.4a–d.

The work area and accessibility characteristics of the Spine robot (Fig. 10.4b) discussed in Section 1.8 are presented below.

Compared with many other industrial robots, a Spine robot requires very little space for installation. It requires no more room than a human being. The dimensions of the base plate are 0.7 × 0.4 m and there are no counterweights which require space behind the robot. It is mounted as standard on the floor. The considerable range of the robot means that the arm can work within a sphere of 50 m^2 and reaches 4 m in an upright position. The arm can rotate repeatedly without needing to go back to a zero point. The robot can also work on several production lines at the same time.

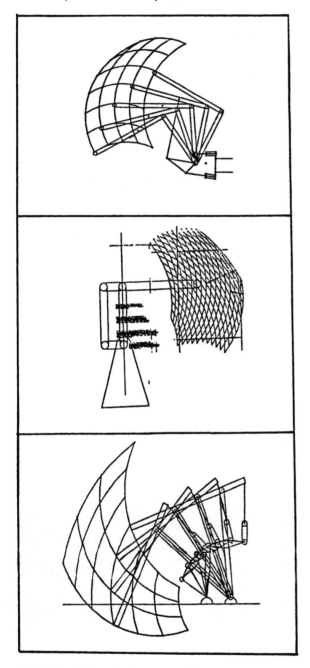

FIGURE 10.2   Workspace of linkage robots generated by a computer.

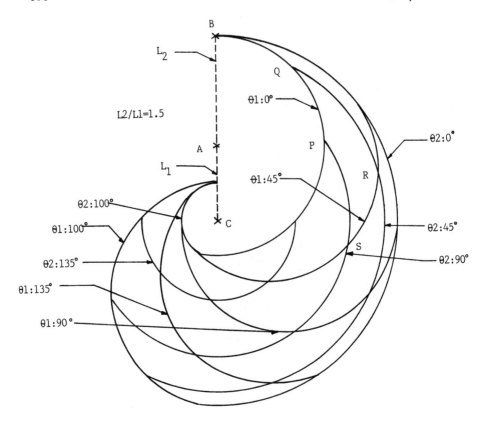

FIGURE 10.3 Design chart for a two-link robot developed on a computer graphics terminal.

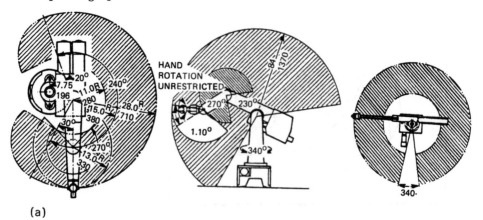

(a)

FIGURE 10.4 (a) Work area of Pickomatic robots (courtesy of Pickomatic Systems). (b) Workspace and range of Spine robots;

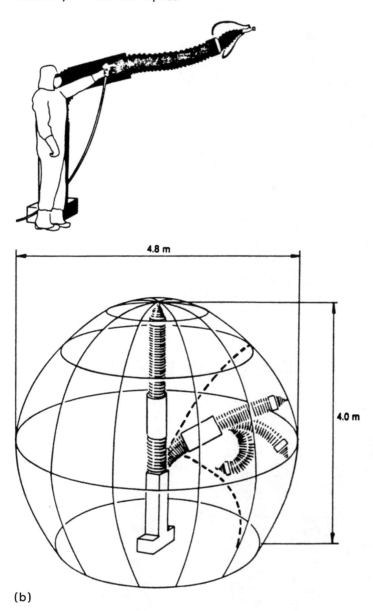

(b)

FIGURE 10.4 (Continued) (c) Accessibility test on a double metal cabinet to be painted by a Spine robot; (d) Accessibility of a robot is important in applications such as rustproofing operations (b—d courtesy of Spine Robotics Corp.).

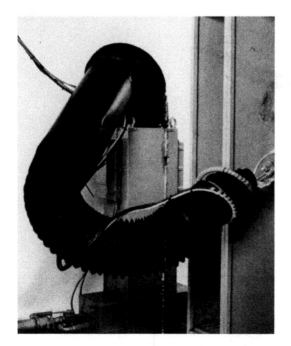

(c)

(d)

FIGURE 10.4 (Continued)

10.2   ANALYSIS OF ROBOT WORKSPACE

Two-Link Robot

Consider the plot of the accessible area of a revolute jointed robot arm with two links shown in Figure 10.5. The coordinates of the hand (x,y) are given by

$$x = \ell_1 \sin \theta_1 + \ell_2 \sin(\theta_1 + \theta_2) \tag{10.1}$$

$$y = \ell_1 \cos \theta_1 + \ell_2 \cos(\theta_1 + \theta_2) \tag{10.2}$$

From these equations one can obtain

$$(x - \ell_1 \sin \theta_1)^2 + (y - \ell_1 \cos \theta_1)^2 = \ell_2^2 \tag{10.3}$$

and

$$x^2 + y^2 = \ell_1^2 + \ell_2^2 + 2\ell_1 \ell_2 \cos \theta_2 \tag{10.4}$$

For a given (x,y) location of the hand the joint angles θ_1 and θ_2 can be solved by using Equations 10.3 and 10.4. This is called the inverse kinematic solution and will be dealt with in detail in a later chapter. Expanding equation 10.3 and simplifying yields

$$x \sin \theta_1 + y \cos \theta_1 = \frac{x^2 + y^2 + \ell_1^2 - \ell_2^2}{2\ell_1}$$

This trigonometric equation can be solved for θ_1:

$$\theta_1 = \cos^{-1} \frac{y}{\sqrt{x^2 + y^2}} - \cos^{-1} \frac{x^2 + y^2 + \ell_1^2 - \ell_2^2}{2\ell_1 \sqrt{x^2 + y^2}}$$

The angle θ_2 is obtained by solving equation 10.4 directly;

$$\theta_2 = \cos^{-1} \frac{x^2 + y^2 - \ell_1^2 - \ell_2^2}{2\ell_1 \ell_2}$$

The boundary of the workarea may be analyzed as follows. The boundary CC_1 is obtained by fixing θ_2 at its minimum value and rotating θ_1 through its range. With θ_1 at the maximum setting, C_1C_2 is obtained by rotation of link 2 through t²e range of θ_2. The circular arc C_2C_3 represents the locus when θ_1 is rotated at a maximum setting θ_2. The fourth curve C_3C is the path obtained by rotating through the range of θ_2 when θ_1 is set at its minimum value (Figure 10.5).

These curves can be represented algebraically as follows:

$$CC_1: \quad x^2 + y^2 = \ell_1^2 + \ell_2^2 + 2\ell_1\ell_2 \cos \theta_{2min}$$

$$C_2C_3: \quad x^2 + y^2 = \ell_1^2 + \ell_2 + 2\ell_1\ell_2 \cos(\theta_{2min} + \theta_{2R})$$

$$C_3C: \quad (x - \ell_1 \sin \theta_{1min})^2 + (y - \ell_1 \cos \theta_{1min})^2 = \ell_2^2$$

$$C_1C_2: \quad [x - \ell_1 \sin(\theta_{1min} + \theta_{1R})]^2 + [y - \ell_1 \cos(\theta_{1min} + \theta_{1R})]^2$$

$$= \ell_2^2$$

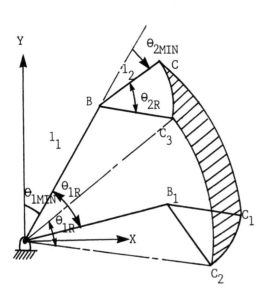

FIGURE 10.5   Analysis of the work area of a two-link robot.

An analytical expression for the area can be derived by considering the segments as follows:

$$\text{area} = ABCC_1C_2A - AC_3C_2A - ABCC_3A$$

$$= (ABCA + ACC_1A + AC_1C_2A) - (AC_3C_2A + ABCC_3A)$$

$$= ACC_1A - AC_3C_2A$$

$$= \frac{1}{2}\,\theta_{1R}\,\{(\ell_1^2 + \ell_2^2 + 2\ell_1\ell_1\,\cos\theta_{2min})$$

$$- [\ell_1^2 + \ell_2^2 + 2\ell_1\ell_2\,\cos(\theta_{2min} + \theta_{2R})]\}$$

$$= \theta_{1R}\ell_1\ell_2(\cos\theta_{2min} - \cos\theta_{2max})$$

The preceding analytical derivations for the work area of a two-link robot can be used in synthesis problems. Consider the synthesis of a two-link robot to reach a set of specified working positions $P_i(x_i,y_i)$, $i = 1,2, \ldots , n$, in the first quadrant. In such synthesis, it is required to determine the link lengths and the location of the base in the workspace so that a minimum work area is obtained. The minimum work area may be obtained by a computer grid search with varying base location coordinates.

For each of the design points P_i the maximum and minimum values of θ_1 and θ_2 from θ_{1i} and θ_{2i} are given by

$$\theta_{1i} = \cos^{-1}\frac{y_i - y_0}{\sqrt{(x_i - x_0)^2 + (y_i - y_0)^2}}$$

$$- \cos^{-1}\frac{(x - x_0)^2 + (y_i - y_0)^2 + \ell_1^2 - \ell_2^2}{2\ell_1\sqrt{(x_i - x_0)^2 + (y_i - y_0)^2}}$$

$$\theta_{2i} = \cos^{-1}\frac{(x_i - x_0)^2 + (y_i - y_0)^2 - (\ell_1^2 + \ell_2^2)}{2\ell_1\ell_2}$$

The work area for each chosen grid point for the base is then computed. The optimum point is the one which gives the minimum area.

Three-Link Robot

For a three-link robot, the total primary work area is obtained by rotating the work area of the outer two links through the range of θ_1, as shown in Figure 10.6. The work areas of typical industrial robots are shown in Figures 10.7 to 10.9.

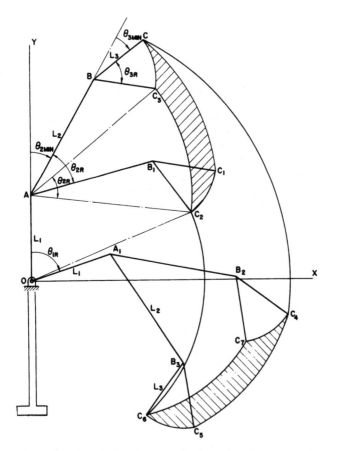

FIGURE 10.6   Construction of the work area of a three-link robot.

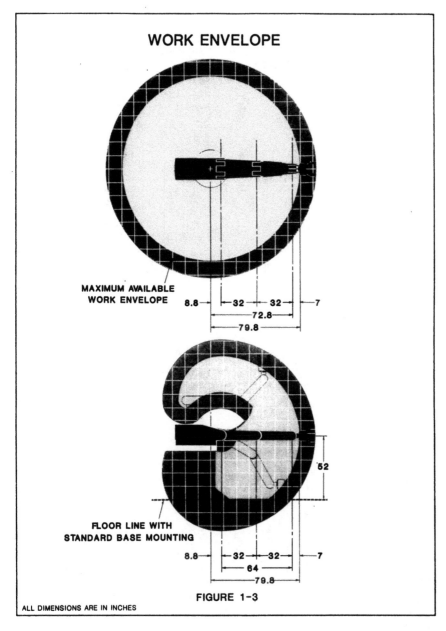

WORK ENVELOPE

MAXIMUM AVAILABLE
WORK ENVELOPE

8.8 — |← 32 →|← 32 →| — 7
— 72.8 —
— 79.8 —

FLOOR LINE WITH
STANDARD BASE MOUNTING

8.8→ |← 32 →|← 32 →| — 7
— 64 —
— 79.8 —

52

FIGURE 1-3

ALL DIMENSIONS ARE IN INCHES

FIGURE 10.7   Work envelope of a robot.   (Courtesy of International Robomation/Intelligence.)

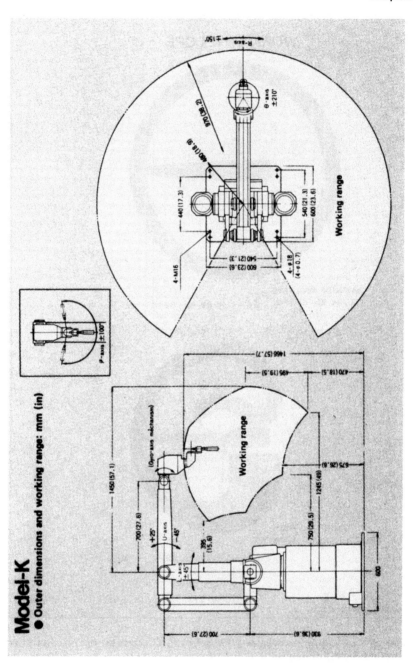

FIGURE 10.8   Geometry of robot workspace.   (Courtesy of Merrick Engineering, Inc.)

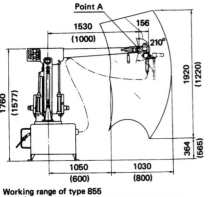

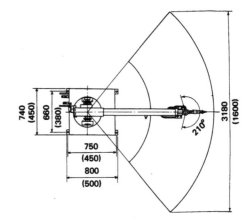

Working range of type 855

Figures in parentheses indicate range of type 705
(Area of point A movement)

FIGURE 10.9   Geometry of robot workspace.   (Courtesy of Tokico America, Inc.)

The equations of the boundaries are:

CC_4:

$$x^2 + y^2 = [\ell_1 + \ell_2 \cos \theta_{2min} + \ell_3 \cos(\theta_{2min} + \theta_{3min})]^2$$

$$+ [\ell_2 \sin \theta_{2min} + \ell_3 \sin(\theta_{2min} + \theta_{3min})]^2$$

$$= \ell_1^2 + \ell_2^2 + \ell_3^2 + 2\ell_1\ell_2 \cos \theta_{2min}$$

$$+ 2\ell_1\ell_3 \cos(\theta_{2min} + \theta_{3min}) + 2\ell_2\ell_3 \cos \theta_{3min}$$

C_4C_5:

$$(x - \ell_1 \sin \theta_{1max})^2 + (y - \ell_1 \cos \theta_{1max})^2$$

$$= \ell_2^2 + \ell_3^2 + 2\ell_2\ell_3 \cos \theta_{3max}$$

C_5C_6:

$$[x - \ell_1 \sin \theta_{1max} - \ell_2 \sin(\theta_{1max} + \theta_{2max})]^2$$

$$+ [y - \ell_1 \cos \theta_{1max} - \ell_2 \cos(\theta_{1max} + \theta_{2max})]^2 = \ell_3^2$$

C_6C_2:

$$x^2y^2 = [\ell_1 \sin \theta_{1max} + \ell_2 \sin \theta_{2max} + \ell_3 \sin \theta_{3max}]^2$$

$$+ [\ell_1 \cos \theta_{1max} + \ell_2 \cos \theta_{2max} + \ell_3 \cos \theta_{3max}]^2$$

C_2C_3:

$$x^2 + (y - \ell_1)^2 = \ell_2^2 + \ell_3^2 + 2\ell_2\ell_3 \cos \theta_{3max}$$

C_3C:

$$(x - \ell_2 \sin \theta_{2min})^2 + (y - \ell_1 - \ell_2 \cos \theta_{2min})^2 = \ell_3^2$$

The work area is

$$\text{area} = \text{sector } OCC_4O + \text{sector } A_1C_4C_5$$

$$- \text{sector } AC_3C_2 - \text{sector } OC_2C_6$$

The distances needed to evaluate the areas are

$$(OC)^2 = [\ell_1 + \ell_2 \cos \theta_{2min} + \ell_3 \cos(\theta_{2min} + \theta_{3min})]^2$$

$$+ [\ell_2 \sin \theta_{2min} + \ell_3 \sin(\theta_{2min} + \theta_{3min})]^2$$

$$= \ell_1^2 + \ell_2^2 + \ell_3^2 + 2\ell_1\ell_2 \cos \theta_{2min}$$

$$+ 2\ell_1\ell_3 \cos(\theta_{2min} + \theta_{3min}) + 2\ell_2\ell_3 \cos \theta_{3min}$$

$$(A_1C_4)^2 = (AC)^2$$

$$= \ell_2^2 + \ell_3^2 + 2\ell_2\ell_3 \cos \theta_{3min}$$

$$(AC_3)^2 = \ell_2^2 + \ell_3^2 + 2\ell_2\ell_3 \cos \theta_{3max}$$

$$(OC_2)^2 = \ell_1^2 + \ell_2^2 + \ell_3^2 + 2\ell_1\ell_2 \cos \theta_{2max}$$

$$+ 2\ell_1\ell_3 \cos(\theta_{2max} + \theta_{3max}) + 2\ell_2\ell_3 \cos \theta_{3max}$$

The areas of the sectors are given by

$$OCC_4O = \frac{\theta_{1R}}{2}(OC)^2$$

$$A_1C_4C_5 = \frac{\theta_{2R}}{2}(AC)^2$$

$$AC_3C_2 = \frac{\theta_{2R}}{2}(AC_3)^2$$

$$OC_2C_6 = \frac{\theta_{1R}}{2}(OC_2)^2$$

Substituting the distances from page 346 and simplifying the result, one obtains the final expression for the work area as

$$
\begin{aligned}
\text{area} = \;\theta_{1R}\{ \ell_1\ell_2(\cos\theta_{2min} &- \cos\theta_{2max}) \\
+\; \ell_2\ell_3(\cos\theta_{3min} &- \cos\theta_{3max}) \\
+\; \ell_1\ell_3[\cos(\theta_{2min} + \theta_{3min}) &- \cos(\theta_{2max} + \theta_{3max})]\} \\
+\; \theta_{2R}\ell_2\ell_3(\cos\theta_{3min} &- \cos\theta_{3max})
\end{aligned}
$$

This expression may be reduced to the work area of a two-link robot by putting ℓ_3 equal to 0. Thus, for a two-link robot the work area is

$$\text{area} = \theta_{1R}\ell_1\ell_2(\cos\theta_{2min} - \cos\theta_{2max})$$

10.3   EXTREME DISTANCES

The extreme distances of manipulator joints and hand provide useful information for the workspace design of robot arms. Two different extreme distances may be defined. The first is the extreme mutual perpendicular distance between the first joint axis and the hand axis. This distance provides a design criterion particularly for robots with prismatic joints. For revolute jointed manipulators this distance is used for describing the effect of the joint variables on the position as well as the orientation of the hand. A further use of this distance is in the force and torque analysis of robot manipulators.

The second type of extreme distance is that between a base point and the center of the hand. The base point is conveniently chosen at the center of the first joint in the arm. The extreme perpendicular distance between the center point of the hand and the first joint axis forms a third type of extreme distance. For a robot located centrally to feed workpieces, this type of extreme distance is an important design parameter.

For robot arms with up to three links, trigonometric methods may be used to derive these extreme distances. For robots with more than three links, very complex mathematical expressions are obtained because of the nonlinear and lengthy functions of the joint variables.

Two-Link Robot Arm

The vector from O to H can be written as

$$\overline{\gamma}_2 = d_1 \hat{k} + a_1 \hat{i}_1 + d_2 \hat{k}_1 + a_2 \hat{i}_2 \tag{10.5}$$

The unit vectors appearing in this expression can be reduced to $(\hat{i}, \hat{j}, \hat{k})$ coordinates by transformation. By inspection, the transformation equations are

$$\hat{i}_1 = C\theta_1 \hat{i} + S\theta_1 \hat{j}$$

$$\hat{j}_1 = -S\theta_1 C\alpha_1 \hat{i} + C\theta_1 C\alpha_1 \hat{j} + S\alpha_1 \hat{k}$$

$$\hat{k}_i = S\theta_1 S\alpha_1 \hat{i} - C\theta_1 S\alpha_1 \hat{j} + C\alpha_1 \hat{k}$$

$$\hat{i}_2 = \hat{i}_1 C\theta_2 + \hat{j}_1 S\theta_2$$

$$= (C\theta_2 C\theta_1 - S\theta_1 S\theta_2 C\alpha_1)\hat{i} + (C\theta_2 S\theta_1 + C\theta_1 S\theta_2 C\alpha_1)\hat{j}$$

$$+ S\alpha_1 S\theta_2 \hat{k} \tag{10.6}$$

Substituting Equation 10.6 into Equation 10.5 yields the position vector:

$$\overline{\gamma}_2 = [a_1 C\theta_1 + d_2 S\theta_1 S\alpha_1 + a_2(C\theta_2 C\theta_1 - S\theta_1 S\theta_2 C\alpha_1)]\hat{i}$$

$$+ [a_1 S\theta_1 - d_2 C\theta_1 S\alpha_1 + a_2(C\theta_2 S\theta_1 + C\theta_1 S\theta_2 C\alpha_1)]\hat{j}$$

$$+ (d_1 + d_2 C\alpha_1 + a_2 S\alpha_1 S\theta_2)\hat{k} \tag{10.7}$$

The distance from O to the center of the hand is obtained by taking the self-scalar product of Equation 10.7 and simplifying the resulting expression. The final result is

$$\gamma_2^{\,2} = a_1^{\,2} + d_1^{\,2} + a_2^{\,2} + d_2^{\,2} + 2a_1 a_2 C\theta_2$$
$$+ 2d_1 d_2 C\alpha_1 + 2d_1 a_2 S\alpha_1 S\theta_2 \tag{10.8}$$

The condition for an extreme value of γ_2 is

$$\frac{d(\gamma_2^{\,2})}{d\theta_2} = 0$$

This yields

$$-2a_1 a_2 S\theta_2 + 2d_1 a_2 S\alpha_1 C\theta_2 = 0 \tag{10.9}$$

Solving Equation 10.9, the final result is obtained as

$$\tan\theta_2 = \frac{d_1 S\alpha_1}{a_1}$$

Two values of θ_2 satisfy this equation and give the maximum and minimum distances by substitution into Equation 10.8. Figure 10.10 gives the notation for two-link extreme distances.

Three-Link Robot Arm

The extreme distance conditions for a three-link robot can be obtained by extending the development of the two-link arm to three links (Figure 10.11). The transformations are

$$\begin{bmatrix} \hat{i}_2 \\ \hat{j}_2 \\ \hat{k}_2 \end{bmatrix} = \begin{bmatrix} C\theta_2 & S\theta_2 & 0 \\ -S\theta_2 C\alpha_2 & C\theta_2 C\alpha_2 & S\alpha_2 \\ +S\theta_2 S\alpha_2 & -C\theta S\alpha_2 & C\alpha_2 \end{bmatrix} \begin{bmatrix} i_1 \\ j_1 \\ k_1 \end{bmatrix}$$

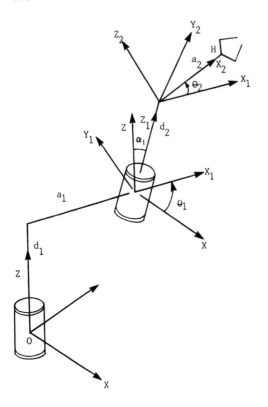

FIGURE 10.10   Notation for developing expression for extreme distance.

$$
\begin{bmatrix} \hat{i}_1 \\[4pt] \hat{j}_1 \\[4pt] \hat{k}_1 \end{bmatrix}
=
\begin{bmatrix}
C\theta_1 & S\theta_1 & 0 \\[4pt]
-S\theta_1 C\alpha_1 & C\theta_1 C\alpha_1 & S\alpha_1 \\[4pt]
S\theta_1 S\alpha_1 & -C\theta_1 S\alpha_1 & C\alpha_1
\end{bmatrix}
\begin{bmatrix} \hat{i} \\[4pt] \hat{j} \\[4pt] \hat{k} \end{bmatrix}
$$

and

$$
\hat{i}_3 = (C\theta_3 C\theta_2 - S\theta_2 S\theta_3 C\alpha_2)\hat{i}_2
$$

$$
+ (C\theta_3 S\theta_2 + C\theta_2 S\theta_3 C\alpha_2)\hat{j}_2
$$

$$
S\alpha_2 S\theta_3 \hat{k}
$$

The position vector of point H in the hand is

$$\bar{\gamma}_3 = d_1 \hat{k} + a_1 \hat{i}_1 + d_2 \hat{k}_1 + a_2 \hat{i}_2 + d_2 \hat{k}_2 + a_3 \hat{i}_3$$

In expanded form this becomes

$$\gamma_3 = \gamma_{3x} \hat{i} + \gamma_{3y} \hat{j} + \gamma_{3z} \hat{k}$$

where

$$\gamma_{3x} = a_1 C\theta_1 + d_2 S\theta_1 S\alpha_1 + a_2 (C\theta_1 C\theta_2 - S\theta_1 S\theta_2 C\alpha_1)$$

$$+ d_3 [C\theta_1 S\theta_2 S\alpha_2 + S\theta_1 (S\alpha_1 C\alpha_2 + C\theta_2 C\alpha_1 S\alpha_2)]$$

$$+ a_3 S\theta_1 [S\theta_3 S\alpha_1 S\alpha_2 - (S\theta_2 C\theta_3 + C\theta_2 S\theta_3 C\alpha_2) C\alpha_1]$$

$$+ a_3 C\theta_1 (C\theta_2 C\theta_3 - S\theta_2 S\theta_3 C\alpha_2)$$

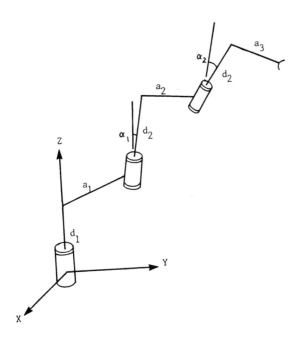

FIGURE 10.11   Notation for determining the extreme distance of a three-link robot.

$$\gamma_{3y} = a_1 S\theta_1 - d_2 C\theta_1 S\alpha_1 + a_2(S\theta_1 C\theta_2 + C\theta_1 S\theta_2 C\alpha_1)$$

$$+ d_3[S\theta_1 S\theta_2 S\alpha_2 - C\theta_1(S\alpha_1 C\alpha_2 + C\theta_2 C\alpha_1 S\alpha_2)]$$

$$- a_3 C\theta_1[S\theta_3 S\alpha_1 S\alpha_2 - (S\theta_2 C\theta_3 + C\theta_2 S\theta_3 C\alpha_2)]$$

$$+ a_3 S\theta_1(C\theta_2 C\theta_3 - S\theta_2 S\theta_3 C\alpha_2)$$

$$\gamma_{3z} = d_1 + d_2 C\alpha_1 + a_2 S\theta_2 S\alpha_1 + d_3(C\theta_1 C\alpha_2 - C\theta_2 S\alpha_1 S\alpha_2)$$

$$+ \varepsilon_3 S\theta_3 C\alpha_1 S\alpha_2 + a_3 S\alpha_1(S\theta_2 C\theta_3 + C\theta_2 S\theta_3 C\alpha_2)$$

The distance γ_3 is given by

$$|\gamma_3|^2 = \bar{\gamma}_3 \cdot \bar{\gamma}_3$$

$$= \gamma_{3x}^2 + \gamma_{3y}^2 + \gamma_{3z}^2$$

Expanding this equation and simplifying, the final form of the distance can be shown to be

$$|\gamma_3|^2 = d_1^2 + a_1^2 + d_2^2 + a_2^2 + d_3^2 + a_3^2$$

$$+ 2d_1[d_2 C\alpha_1 + a_2 S\theta_2 C\alpha_1 + d_3(C\alpha_1 C\alpha_2 - C\theta_2 S\alpha_1 S\alpha_2)]$$

$$+ 2d_1 a_3[S\theta_3 C\alpha_1 S\alpha_2 + S\alpha_1(S\theta_2 C\theta_3 + C\theta_2 S\theta_3 C\alpha_2)]$$

$$+ 2a_1[a_2 C\theta_2 + d_3 S\theta_2 S\alpha_2 + a_3(C\theta_2 C\theta_3 = S\theta_2 S\theta_3 C\alpha_2)]$$

$$+ 2d_2(d_3 C\alpha_2 + a_3 S\theta_3 S\alpha_2) + 2a_2 a_3 C\theta_3$$

The conditions for an extremum are given by

$$\frac{\partial(\gamma_3^2)}{\partial\theta_2} = 0$$

and

$$\frac{\partial(\gamma_3^2)}{\partial\theta_3} = 0$$

These conditions yield, respectively,

$$P_1 S\theta_3 + Q_1 C\theta_3 + R_1 = 0$$

and

$$P_2 S\theta_3 + Q_2 C\theta_3 = 0$$

where

$$P_1 = -d_1 a_3 S\alpha_1 C\alpha_2 S\theta_2 - a_1 a_3 C\alpha_2 C\theta_2$$

$$Q_1 = -a_1 a_3 S\theta_2 + d_1 a_3 S\alpha_1 C\theta_2$$

$$R_1 = S\theta_2(d_1 d_3 S\alpha_1 S\alpha_2 - a_1 a_2) + S\alpha_1 C\theta_2(d_1 a_2 + d_3 a_1)$$

$$P_2 = -d_1 a_3 S\alpha_1 S\theta_2 - a_1 a_3 C\theta_2 - a_2 a_3$$

$$Q_2 = -a_1 a_3 C\alpha_2 S\theta_2 + d_1 a_3 C\alpha_2 C\theta_2$$
$$\qquad + a_3 S\alpha_2(d_1 C\alpha_1 + d_2)$$

Influence of Link Parameters

For any manipulator the total workspace may be obtained by sequentially rotating each axis, starting from the terminal device. The
workspace relative to the last joint axis is a circle, as shown in
Figure 10.12. When this is rotated about the next joint, a torus is
obtained. The common anchor ring shown in Figure 10.12 is the
right circular form of the torus obtained when adjacent joint axes
are orthogonal.

For a general three-joint robot as shown in Figure 10.13 the
work area is influenced by the link parameters a_1, α_1, and d_1. The
effects of a_1, α_1, and d_1 can be studied by writing the rectangular
components of the position of the terminal device in terms of the
link parameters. The contours of the work area can then be determined, typically by an interactive computer with graphics capabilities. If the link lengths a_2 and a_3 are equal, the work areas do
not have any voids and the maximum work area is obtained by setting a_1 to zero. Additional constraints for symmetrical work areas
are set α_1 and α_2 to 0° or 90° and set d_2 and d_3 to zero. The
angle α_2 does not affect the shape of the work area under the conditions $a_1 = 0$, $a_2 = a_3$, and $d_2 = d_3 = 0$. However, increasing α_2
from 0° to 90° increases the voids in the work area. When α_2 is
set at 0° or 180° and the link lengths a_2 and a_3 are equal, there

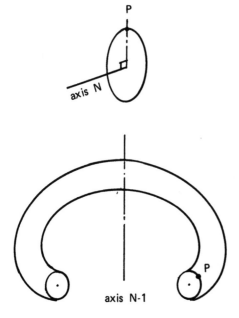

FIGURE 10.12   Generation of circle and torus of robot workspace.

are no voids in the work area for all values of α_1. The first link parameter α_1 will clearly not have any effect on the shape of the work area, but will shift it vertically. The link parameters d_2 and d_3 affect the size of the voids in the work area.

In mechanisms, Grashof's criteria permit the analysis of the conditions under which a planar four-bar linkage has a completely rotatable crank. Locating the dextrous workspace of a manipulator is analogous to Grashof's criteria. For a planar three-link manipulator with the hand rotating about the center point, the closed loop gives rise to a four-bar mechanism. The relative dimensions of this planar analog mechanism for complete rotatability of the hand is derived directly from Grashof's laws.

In the planar analog, complete hand rotatability is generally obtained by using a short hand. In the spatial analog, a short hand for complete rotatability means that the normal distance from the center point to the last axis must be very short. That is, the terminal device should be very close to the approach axis of the hand. A manipulator of arbitrary geometry does not necessarily have a dextrous workspace. Generally a short hand favors tme existence of a dextrous workspace.

One class of manipulator geometries which can be shown to have dextrous workspaces is the anthropomorphic structure without offsets. In this manipulator the base (waist) and shoulder axes are orthogonal and the hand orientation axes intersect each other at a point and are also orthogonal. If a center point is chosen to coincide with the point of intersection of the axes of the hand, the workspace takes the form shown in Figure 10.14. At any point in the workspace, complete rotation about each of the hand axes is possible. The dextrous workspace becomes identical to the accessible workspace. Another possibility is locating the reference point at the end of the terminal device. Consider a worst configuration in which the reference point axis is normal to the plane of the first two lines (forearm and elbow). The boundaries of the workspaces are shown in Figure 10.15. Contours c and d correspond to the folded-in hand position and define the boundaries of the dextrous workspace. For small hand sizes, a large percentage of the workspace is a dextrous workspace. As the hand size increases the dextrous workspace decreases and eventually vanishes. At the

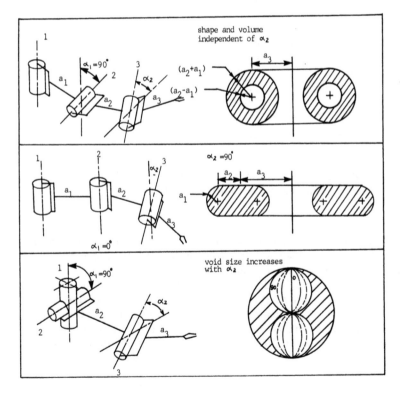

FIGURE 10.13 Variation of the shape of the cross section of the torus with the link parameters.

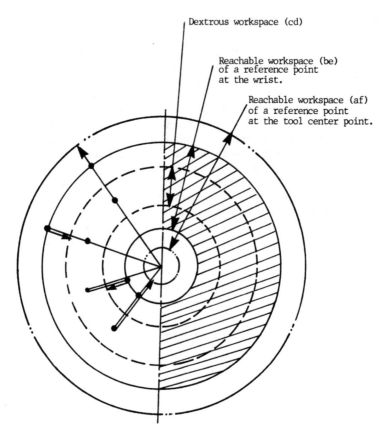

Dextrous workspace (cd)

Reachable workspace (be)
of a reference point
at the wrist.

Reachable workspace (af)
of a reference point
at the tool center point.

FIGURE 10.14 Contours defining the dextrous and reachable work-spaces of a three-link revolute-jointed robot.

same time the total workspace increases as the hand size increases. When the hand size exceeds a certain value, the workspace may undergo a major geometric change.

The analysis above, which is based on the positioning of a point fixed in the hand of the manipulator, can be extended to deal with the influence of the manipulator link parameters on the position and orientation of a line fixed to the manipulator. Line geometry, as opposed to the use of points, is a convenient technique for robots with prismatic joints. In such cases, typically, no single point along the sliding axis is of interest, but the axis itself must be studied. The method of analysis is based on Plucker coordinates and is beyond the scope of this text.

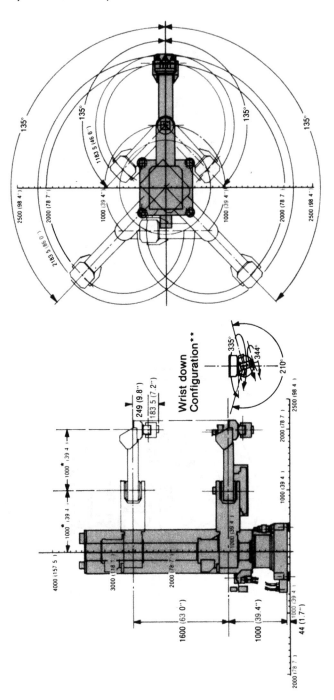

FIGURE 10.15 Workspace geometry of Cybotech robots. (Courtesy of Cybotech Corporation.)

358

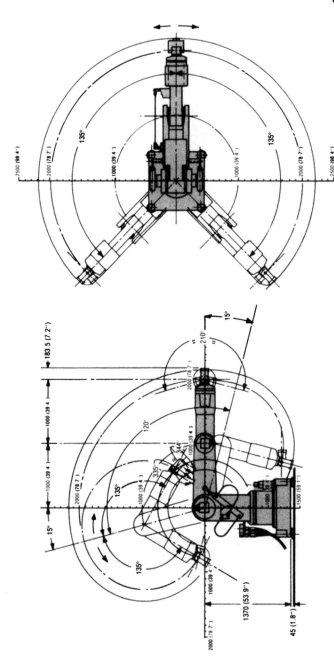

FIGURE 10.15   (Continued)

REVIEW QUESTIONS

1. Trace out the work envelopes of a number of industrial robots using drawing instruments. Write a computer graphics program to determine the locus of points defining the work envelopes. Compare the plots you obtain with manufacturer's data and comment on any differences you note.
2. Study a number of applications of industrial robots and discuss the workspaces required.
3. Using the equations developed in this chapter, obtain suitable values for the work area and extensive reach for the given robots whose dimensions are known.
4. Write a general computer program with subroutines to obtain the numerical values in problem 3.

FURTHER READING

Denavit, J. and Hartenburg, R. S., "A Kinematic Notation for Lower Pair Mechanisms Based on Matrices," *J. Appl. Mech.*, Vol. 22, *Trans. ASME*, Vol. 77, 1955, pp. 215–221.

Duffy, J., *Analysis of Mechanisms and Robot Manipulators*, Wiley, New York, 1980.

Duffy, J. and Rooney, J., "On the Closures of Spatial Mechanisms," ASME Paper No. 72-Mech-77, Mechanisms Conference, 1972.

Fichter, E. F. and Hunt, K. H., "The Fecund Torus, Its Bitangent Circles and Derived Linkages," *Mech. Mach. Theory*, Vol. 10, 1975, pp. 167–176.

Gupta, K. C. and Roth, B., "Design Consideration for Manipulator Workspace," ASME Paper No. 810330, 1981.

Kumar, A. and Waldron, K. J., "The Dextrous Workspace," ASME Paper No. 80-DET-108, 1980.

Kumar, A. and Waldron, K. J., "The Workspace of a Mechanical Manipulator," *ASME J. Mech. Des.*, Vol. 103, No. 3, July 1981, pp. 665–672.

Paul, R. P., "Modeling, Trajectory Calculation and Servoing of a Computer Controlled Arm," Stanford Intelligence Laboratory, Stanford University, AIM 177, 1972.

Roth, B., "Robots," *Appl. Mech. Rev.*, Vol. 31, No. 11, 1978, pp. 1511–1519.

Sugimoto, K. and Duffy, J., "Determination of Extreme Distances of A Robot Hand—Part 1, A General Theory," ASME Paper No. 80-DET-57, presented at ASME 16th Mechanisms Conference, California, Sept. 28–Oct. 1, 1980.

Sugimoto, K. and Duffy, J., "Determination of Extreme Distances
    of A Robot Hand—Part 1, A General Theory,"ASME Paper No.
    80-DET-58, presented at ASME 16th Mechanisms Conference,
    California, Sept. 28–Oct. 1, 1980.
Tesar, D., "A Review of Modeling, Control and Design of Manipula-
    tors in Terms of the Planar N DOF System," unpublished notes,
    University of Florida, Gainesville, Jan. 1978.
Tsai, Y. C., "Synthesis of Robots/Manipulators for a Prescribed
    Working Space," Ph.D. dissertation, Oklahoma State University,
    1981.
Tsai, Y. C. and Soni, A. H., "Accessible Region and Synthesis of
    Robot Arms," ASME Paper No. 80-DET-101, presented at ASME
    16th Mechanisms Conference, California, Sept. 28–Oct. 1, 1980.
Vertut, J., "Contribution to Analyze Manipulator Morphology Cover-
    age and Dexterity," On the Theory and Practice of Manipulators,
    Vol. 1, Springer-Verlag, New York, 1974, pp. 277–289.

11

Kinematics
and Trajectory Generation

11.1 KINEMATICS EQUATIONS

The position and orientation of the end effector of a robot are frequently expressed by the vectors p and (n, o, a) shown in Figure 11.1 The components of these vectors are given in terms of a fixed Cartesian coordinate frame at the base of the robot. The approach vector a is the unit vector in the direction in which the end effector will normally approach an object. The slide vector o lies along the fingertip-to-fingertip line of a two-fingered gripper. The third orientation vector is a normal vector n which completes a right-hand triad of unit vectors. This can be expressed by the vector product relation

$$n = o \times a$$

There are other coordinate variables for determining the position and orientation of the end effector. Alternative specifications of position can be expressed by cylindrical and spherical coordinates. Alternative orientation specifications are the Euler and RPY (roll-pitch-yaw) transformations. We will now state these alternative transformations.

In cylindrical coordinates (r, θ, z) the position of the end effector is

$$p_x = r \cos \theta$$

$$p_y = r \sin \theta \qquad (11.1)$$

$$p_z = z$$

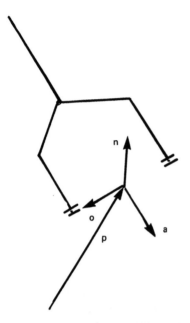

FIGURE 11.1   Specification of basic position and orientation vectors of a robot's end effector.

In spherical coordinates (R, θ, ϕ) the Cartesian components can be shown to be

$$p_x = R \cos \theta \sin \phi$$

$$p_y = R \sin \theta \sin \phi \qquad\qquad (11.2)$$

$$p_z = R \cos \phi$$

The Euler transformation is defined by a sequence of rotations ϕ, θ, ψ about local z, y, x axes. We may therefore write

$$\text{Euler}(\phi, \theta, \psi) = \text{Rot}(z, \phi) * \text{Rot}(y, \theta) * \text{Rot}(z, \psi) \qquad (11.3)$$

This equation means that the Euler transformation is obtained by a rotation ϕ about the z axis, then a rotation θ about the current y axis, and finally a rotation ψ about the current z axis, z''. The sequence of rotations may also be interpreted in the reverse order as rotations in base coordinates. In this interpretation one starts with a rotation ψ about the z axis, followed by a rotation θ about the fixed y axis and then a final rotation ϕ about the fixed z axis.

Expanding equation 11.3, the Euler transformation can be obtained as follows:

$$\text{Euler}(\phi,\theta,\psi) = \text{Rot}(z,\phi) * \begin{bmatrix} c\theta & 0 & s\theta \\ 0 & 1 & 0 \\ -s\theta & 0 & c\theta \end{bmatrix} * \begin{bmatrix} c\psi & -s\psi & 0 \\ s\psi & c\psi & 0 \\ 0 & 0 & 1 \end{bmatrix}$$

$$= \begin{bmatrix} c\phi & -s\phi & 0 \\ s\phi & c\phi & 0 \\ 0 & 0 & 1 \end{bmatrix} * \begin{bmatrix} c\theta c\psi & c\theta s\psi & s\theta \\ s\psi & c\psi & 0 \\ -s\theta c\psi & -s\theta c\psi & c\theta \end{bmatrix}$$

$$= \begin{bmatrix} c\phi c\theta c\psi - s\phi s\psi & -c\phi c\theta s\psi - s\phi c\psi & c\phi s\theta \\ s\phi c\theta c\psi + c\phi s\psi & -s\phi c\theta s\psi + c\phi c\psi & s\phi s\theta \\ -s\theta c\psi & s\theta s\psi & c\theta \end{bmatrix} \quad (11.4)$$

The orientation vectors [n o a] of the end effector in terms of the Euler coordinates ϕ, θ, ψ become

$$n_x = c\phi c\theta c\psi - s\phi s\psi$$

$$n_y = s\phi c\theta c\psi + c\phi s\psi \quad\quad\quad\quad\quad (11.5)$$

$$n_z = -s\theta c\psi$$

$$o_x = -c\phi c\theta s\psi - s\phi c\psi$$

$$o_y = -s\phi c\theta s\psi + c\phi c\psi$$

$$o_z = s\theta s\psi$$

$$a_x = c\phi s\theta$$

$$a_y = s\phi s\theta$$

$$a_z = c\theta \quad\quad\quad\quad\quad\quad\quad\quad\quad (11.6)$$

Roll, pitch, and yaw, which are frequently used in the motion of ships and airplanes, are another set of rotations for describing the orientation of the end effector. The order of rotations is specified by

$$RPY(\phi,\theta,\psi) = Rot(z,\phi) * Rot(y,\theta) * Rot(x,\psi) \tag{11.7}$$

The RPY transformation matrix is derived as

$$RPY(\phi,\theta,\psi) = Rot(z,\phi) * \begin{bmatrix} c\theta & 0 & s\theta \\ 0 & 1 & 0 \\ -s\theta & 0 & c\theta \end{bmatrix} \begin{bmatrix} 1 & 0 & 0 \\ 0 & c\psi & -s\psi \\ 0 & s\psi & c\psi \end{bmatrix}$$

$$= \begin{bmatrix} c\phi & -s\phi & 0 \\ s\phi & c\phi & 0 \\ 0 & 0 & 1 \end{bmatrix} \begin{bmatrix} c\theta & s\theta s\psi & s\theta c\psi \\ 0 & c\psi & -s\psi \\ -s\theta & c\theta s\psi & c\theta c\psi \end{bmatrix}$$

$$= \begin{bmatrix} c\phi c\theta & c\phi s\theta s\psi - s\phi c\psi & c\phi s\theta c\psi + s\phi s\psi \\ s\phi c\theta & s\phi s\theta s\psi + c\phi c\psi & s\phi s\theta c\psi - c\phi s\psi \\ -s\theta & c\theta s\psi & c\theta c\psi \end{bmatrix} \tag{11.8}$$

From this development we obtain the orientation in terms of the RPY rotations as

$$n_x = c\phi c\theta$$

$$n_y = s\phi c\theta$$

$$n_z = -s\theta$$

$$o_x = c\phi s\theta s\psi - s\phi c\psi$$

$$o_y = s\phi s\theta s\psi + c\phi c\psi$$

$$o_z = c\theta s\psi$$

$$a_x = c\phi s\theta c\psi + s\phi s\psi$$

$$a_y = s\phi s\theta c\psi - c\phi s\psi$$

$$a_z = c\theta c\psi \tag{11.9}$$

11.2 INVERSE KINEMATICS

The inverse kinematics problem arises when the position and orientation of the end effector of a robot are known and the corresponding joint coordinates are required. In this section we will use direct
trigonometric methods to determine the inverse kinematic solutions.
We will present the results for position and orientation coordinates
separately.

Inverse kinematic solutions depend on the direct solution of trigonometric equations of the form

$$A \cos \theta + B \sin \theta = C \tag{11.10}$$

Case 1: $C = 0$, so $\tan \theta = -A/B$. Then

$$\theta = \tan^{-1}\left(-\frac{A}{B}\right) \tag{11.11}$$

or

$$\theta = \pi + \tan^{-1}\left(-\frac{A}{B}\right)$$

Case 2: $C \neq 0$

$$\frac{A}{\sqrt{A^2 + B^2}} \cos \theta + \frac{B \sin \theta}{\sqrt{A^2 + B^2}} = \frac{C}{\sqrt{A^2 + B^2}} \tag{11.12}$$

or

$$\sin(\theta + \phi) = D$$

where $D = C/\sqrt{A^2 + B^2}$ and $\phi = \tan^{-1}(A/B)$. The solution for θ is
thus

$$\theta = \tan^{-1}\left(\frac{D}{\sqrt{1 - D^2}}\right) - \phi. \tag{11.13}$$

For a two-link revolute jointed robot with θ measured from the
horizontal

$$p_x = \ell_1 \cos \theta_1 + \ell_2 \cos(\theta_1 + \theta_2)$$

$$p_y = \ell_1 \sin \theta_1 + \ell_2 \sin(\theta_1 + \theta_2) \tag{11.14}$$

From these equations,

$$p_x \cos \theta_1 + p_y \sin \theta_1 = C$$

where $C = (p_x{}^2 + p_y{}^2 + \ell_1{}^2 - \ell_2{}^2)/2\ell_1$. The solution of this trigonometric equation yields

$$\theta_1 = \tan^{-1}\left(\frac{D}{\sqrt{1 - D^2}}\right) - \tan^{-1}\left(\frac{p_y}{p_x}\right) \qquad (11.15)$$

where $D = C/\sqrt{p_x{}^2 + p_y{}^2}$

Once θ_1 has been obtained, θ_2 can be found as:

$$\theta_2 = \tan^{-1}\left\{\frac{p_y - \ell_1 \sin \theta_1}{p_x - \ell_1 \cos \theta_1}\right\} - \theta_1$$

For a three-link robot

$$p_x = \cos \theta_1[\ell_2 \sin \theta_2 + \ell_3 \sin(\theta_2 + \theta_3)] \qquad (11.16)$$

$$p_y = \sin \theta_1[\ell_2 \sin \theta_2 + \ell_3 \sin(\theta_2 + \theta_3)] \qquad (11.17)$$

$$p_z = \ell_2 \cos \theta_2 + \ell_3 \cos(\theta_2 + \theta_3) \qquad (11.18)$$

The inverse kinematic solution is obtained as follows. From Equations 11.16 and 11.17

$$\tan \theta_1 = \frac{p_y}{p_x}$$

With θ_1 known, Equations 11.17 and 11.18 yield

$$\frac{p_y}{\sin \theta_1} = \ell_2 \sin \theta_2 + \ell_3 \sin(\theta_2 + \theta_3) \qquad (11.19)$$

$$p_z = \ell_2 \cos \theta_2 + \ell_3 \cos(\theta_2 + \theta_3) \qquad (11.20)$$

Equations 11.19 and 11.20 are of the same form as Equation 11.14. Proceeding in a similar fashion,

$$\theta_2 = \tan \cfrac{D}{\sqrt{1 - D^2}} - \tan^{-1}\left(\frac{P_z \sin \theta_1}{P_y}\right)$$

where

$$D = \cfrac{C}{\sqrt{(p_y/\sin \theta_1)^2 + p_z^2}}$$

$$C = \cfrac{(p_y/\sin \theta_1)^2 + p_z^2 + \ell_2^2 - \ell_3^2}{2\ell_2}$$

Finally,

$$\theta_3 = \tan^{-1}\left(\frac{p_y/\sin \theta_1 - \ell_2 \sin \theta_2}{p_z - \ell_2 \cos \theta_2}\right)$$

Given p_x, p_y, and p_z, the corresponding cylindrical coordinates can be found from

$$p_x = d_3 \cos \theta_1 \qquad p_y = d_3 \sin \theta_1 \qquad p_z = d_2$$

as

$$\theta_1 = \tan^{-1}\left(\frac{p_y}{p_x}\right)$$

$$d_2 = p_z \tag{11.21}$$

and

$$d_3 = \sqrt{p_x^2 + p_y^2}$$

The inverse solution for position specifications given in spherical coordinates is derived from the equations

$$p_x = d_3 \cos \theta_1 \sin \theta_2$$

$$p_y = d_3 \sin \theta_1 \sin \theta_2$$

$$p_z = d_3 \cos \theta_2$$

as

$$d_3 = \sqrt{p_x^2 + p_y^2 + p_z^2}$$

$$\theta_2 = \tan^{-1}\left(\frac{\sqrt{d_3^2 - p_z^2}}{p_z}\right) \qquad (11.22)$$

$$\theta_1 = \tan^{-1}\left(\frac{p_y}{p_x}\right)$$

11.3 DIFFERENTIAL KINEMATICS

The velocity of the end effector of simple robots can be determined by direct differentiation of the kinematic equations. For a two-link robot it has been shown that (θ measured from the vertical)

$$p_x = \ell_1 \sin \theta_1 + \ell_2 \sin(\theta_1 + \theta_2)$$

$$p_y = \ell_1 \cos \theta_1 + \ell_2 \cos(\theta_1 + \theta_2) \qquad (11.23)$$

Differentiating, we obtain

$$\dot{p}_x = \ell_1 \dot{\theta}_1 \cos \theta_1 + \ell_2(\dot{\theta}_1 + \dot{\theta}_2) \cos(\theta_1 + \theta_2)$$

$$\dot{p}_y = -\ell_1 \dot{\theta}_1 \sin \theta_1 - \ell_2(\dot{\theta}_1 + \dot{\theta}_2) \sin(\theta_1 + \theta_2) \qquad (11.24)$$

This can be written in the Jacobian form

$$\begin{bmatrix} \dot{p}_x \\ \dot{p}_y \end{bmatrix} = [J] \begin{bmatrix} \dot{\theta}_1 \\ \dot{\theta}_2 \end{bmatrix} \qquad (11.25)$$

where

$$J = \begin{bmatrix} \ell_1 \cos \theta_1 + \ell_2 \cos(\theta_1 + \theta_2) & \ell_2 \cos(\theta_1 + \theta_2) \\ -\ell_1 \dot{\theta}_1 \sin \theta_1 - \ell_2(\theta_1 + \theta_2) & -\ell_2 \sin(\theta_1 + \theta_2) \end{bmatrix}$$

As a second example, consider the three-link robot whose kinematic equations are given by (θ_1, θ_2, θ_3 are joint variables)

$$P_x = C_1(\ell_2 S_2 + \ell_3 S_{23})$$

$$P_y = S_1(\ell_2 S_2 + \ell_3 S_{23}) \qquad\qquad (11.26)$$

$$P_z = \ell_2 C_2 + \ell_3 C_{23}$$

Where $C_i = \cos\theta_i$, $S_i = \sin\theta_i$, and $C_{ij} = \cos(\theta_i + \theta_j)$. The final velocity components obtained by direct differentiation are given by the vector matrix equation

$$
\begin{bmatrix} \dot{P}_x \\ \dot{P}_y \\ \dot{P}_z \end{bmatrix} =
\begin{bmatrix}
-(\ell_2 S_2 + \ell_3 S_{23})S_1 & (\ell_2 C_2 + \ell_3 C_{23})C_1 & \ell_3 C_{23} C_1 \\
(\ell_2 S_2 + \ell_3 S_{23})C_1 & (\ell_2 C_2 + \ell_3 C_{23})S_1 & \ell_3 C_{23} C_1 \\
0 & -(\ell_2 S_2 + \ell_3 S_{23}) & -\ell_3 S_{23}
\end{bmatrix}
\begin{bmatrix} \dot{\theta}_1 \\ \dot{\theta}_2 \\ \dot{\theta}_3 \end{bmatrix}
$$

$$(11.27)$$

11.4   TRAJECTORY GENERATION SCHEMES

In a general robot task, the end effector is required to move from one point to another in a manner specified by the operator. During these moves the robot must avoid collisions with obstacles, and the end effector may also be required to trace a specified path. Robot path specification is frequently done by requiring the end effector to pass through a number of intermediate points between the initial and final points. These intermediate points are known as way points. There are two methods of moving a robot between the way points: joint interpolated motion and Cartesian motion. Either of these methods may be executed in continuous path mode, in which constant velocity is maintained along the trajectory. The constant velocity causes problems in straight-line motion because of the necessity for an instantaneous acceleration at each point. This means that at the specified way points, the trajectory must be rounded, causing the way point to be missed. On the other hand, in Cartesian motion without continuous path the robot must stop instantaneously at each point. This may be detrimental in some applications such as spray painting and close-tolerance machining. A trajectory generation scheme based on spline functions does not have some of the problems with continuous path motion. Before discussing the concepts used in the trajectory generation schemes of robot motion, we will present the various kinematic schemes that can be used in robot motion.

Point-to-Point Motion

In point-to-point motion each joint is controlled by an individual position servo with all joints moving from position to position independently. There are three ways to control point-to-point motion independently: (1) sequential joint control (PTP-S), (2) uncoordinated joint control (PTP-U), and (3) terminally coordinated joint control (PTP-T).

Sequential Joint (PTP-S) Motion. Sequential joint operation is the activation of one joint at a time while all other axes are immobilized. A single joint may operate more than once in a sequence associated with such a motion. The resulting path of the manipulator end effector will have a zigzag form associated with the motion directions of the manipulator joints. Sequential joint control results in immediate simplifications in the control of an industrial robot. However, sequential control causes longer point-to-point motion times. This type of control will be helpful for highly modularized programmable assemblers in which individual joints can easily be arranged into multijoint manipulators.

Uncoordinated Joint (PTS-U) Motion. In this mode the motions are not coordinated, in sense that one joint having made some fraction of its motion does not imply that all other joints have made the same fractions of their respective motions. When each joint reaches its final position it holds and waits until the joints have completed their motions. Since there is no coordination of motion between joints, the path and velocity of the end effector between points is not easily predicted.

Terminally Coordinated Joint (PTP-T) Motion. This is the most useful type of point-to-point control. The individual joint motions are coordinated so that all joints attain their final positions simultaneously. It is used primarily in applications where only the final position is of interest and the path is not a prime consideration.

Continuous Path Trajectory

This type of control is used where a continuous path of the end effector is of primary importance to the application. Continuous path motions are produced by interpolating each joint control variable from its value to its desired final value. Each joint is moved the minimum amount required to achieve the desired final positions in order to give the robot tool a controlled predictable path. All the joint variables are interpolated to make the joints complete their motions simultaneously, thus giving a coordinated joint motion.

Continuous path techniques can be divided into three categories, based on how much information about the path is used in the motor control calculations (Figure 11.2). The first is the conventional, or

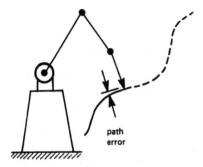

(a) Basic Servo Control

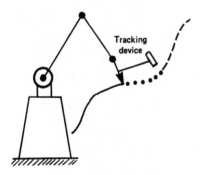

(b) Preview Control

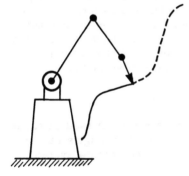

(c) Trajectory Calculation

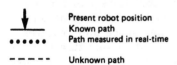

	Present robot position
	Known path
••••••	Path measured in real-time
– – – –	Unknown path

FIGURE 11.2   Three categories of continuous path generation.

servo control, approach. This method uses no information about where
the path goes in the future. The controller may have a stored representation of the path it is to follow, but for determining the drive
signals to the robot's motors all calculations are based on the past
and present path tracking error. This is the control design used in
most of today's industrial robots and process control systems.

The second approach is the preview method, also known as feedforward control. It uses some knowledge about how the path changes
immediately ahead of the robot's current location, in addition to the
past and present tracking error used by the servo controller.

The last category is the path planning or trajectory calculation
approach. Here the computer controller has available a complete
description of the path the manipulator should follow from one point
to another. Using a mathematical model of the arm and its load, the
controller precomputes an acceleration profile for every joint, pre-
dicting the nominal motor signals that should cause the arm to follow
the desired path. This approach has been used in some advanced
research robots to achieve highly accurate coordinated movements at
high speed.

Continuous path motion is actually a point-to-point trajectory
which is sampled on a time base rather than as discretely determined
spatial positions. The result is a smooth continuous motion over a
controlled path, where there is no detectable change in speed from
point to point. The high rate of sampling, typically in the range 60
to 80 Hz, requires many spatial positions to be stored in memory.
Depending on the controller and data storage system used, multiple
programs may be stored in memory for random assessing. In some
applications point-to-point motion is combined with continuous motion.
In such situations, point-to-point motion gets the hand of the robot
to the desired location. Then continuous path takes it through the
precise motions for complex part assembly or similar tasks.

Cartesian Motion Trajectiories

In Cartesian motion generation, the locations are expressed in terms
of the Cartesian coordinates (x,y,z) and orientation angles of the
robot tool relative to a reference frame (world coordinate system)
fixed in the base of the robot.

Cartesian motions are generated by applying an interpolating
function to the Cartesian location of the manipulator's tool tip and
rapidly transforming the interpolated tool tip location to joint commands. The path is prescribed in rectangular space and requires
that the kinematic equations of motion be solved in real time to output continuously to the axis servos. However, the use of Cartesian
coordinates to define locations can be somewhat inaccurate at times
due to the complexity of the computations necessary to convert between the Cartesian coordinates and joint angles via homogeneous
transformations for a general six-axes robot.

When teaching such a system, the computer solves the equations of motion so that the axes are coordinated to give the desired motion, such as straight up in the z-coordinate direction or left in the x-coordinate direction. The operator thinks only in terms of the desired motion of the end effector, not the motion required at the axes. Also, the operator can change the orientation of the end effector about a point without changing the position of the point, or independently change position or orientation of the end effector relative to the tool center point (TCP).

Cartesian motion is useful in full tracking applications requiring synchronization with a moving conveyer using a stationary base robot. In the elementary form of stationary base tracking, the robot is taught the task while the conveyer or part is in motion. During replay of the taught program, the speed of replay and the speed of robot motion are synchronized with the line. This type of tracking system is satisfactory only for point-to-point tasks. Continuous path tasks requiring the TCP to maintain the same velocity relative to the part as when taught cannot use this system unless the conveyer speed is constant without interruption in its motion. For example, when spray painting on an automotive assembly line, if the conveyer stops, the robot cannot stop but must continue to move at the same velocity relative to the part. For applications where constant uninterrupted conveyer speed cannot be ensured, full tracking capability is required to perform the task, independent of conveyer motion (Figure 11.3).

For full tracking and other similar tasks, Cartesian motion control must be extended to the controlled trajectory method.

Controlled Trajectory System

This control mode provides position, velocity, and acceleration control of the robot terminal device along a desired path between programmed points. Also, the method gives coordinated control of the robot joints during the teaching cycle. This feature means that the hand can be positioned in the correct orientation without having to command each joint separately or grasp the robot. Another advantage is that only the end points of a desired path need be programmed. Then in the automatic mode the computer generates the controlled path at the desired velocity and acceleration or deceleration. In teaching robots with controlled path capabilities, Cartesian coordinates are commonly used. A cylindrical coordinate system is also used on some robots. By using a straight-line path between programmed points, the robot hand may be smoothly accelerated and decelerated without centrifugal forces being exerted. This is particularly important when the robot is used for transfer of heavy parts. Since the acceleration and deceleration of the parts are controlled, higher velocities with consequent short cycle times may be used. The reduction of jerky motion with controlled acceleration improves the robot's service life.

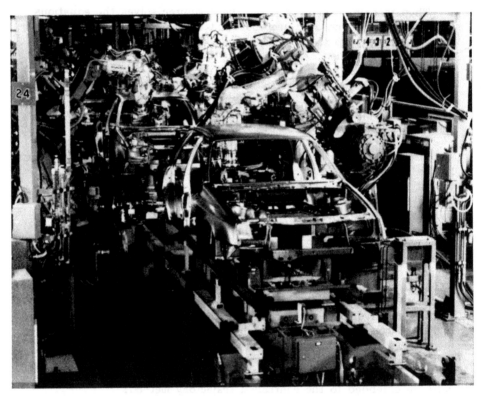

(a)

FIGURE 11.3   Programming features of a robot. (a) Drilling air-
plane parts requires controlled-path straight line motion. (b) Unique
tracking capabilities are required in spot welding car bodies as they
move past on conveyers. (Courtesy of Cincinnati Milacron.)

   In implementing full tracking with the controlled path system, a
continuous real-time measurement of the conveyer or part position
provides a zero shift of the generated path. The measurement is
made along the world x, y, or z axis. This signal is then summed
with the current path position coordinates obtained from the non-
shifted programmed locations. The difference between the rectangular
coordinate position of the hand and the current reading from the
conveyer is then stored in computer memory. With full tracking
capability, the robot operations proceed without interruption and
independent of the conveyor motions.

(b)

FIGURE 11.3 (continued)

Using the controlled trajectory system, a number of robot applications which would be tedious to program in other modes become much easier. In machine loading, for example, programming the intricate moves required to position a large workpiece between centers in a lathe is made easier by the coordinated axes control. In metal inert gas seam welding a straight line between programmed points results in a straighter weld line than could be made manually. The added feature of controlled velocity results in better weld quality and consistency because of the smooth and even motion of the weld tip.

11.5 CONSTRAINT-BASED TRAJECTORIES

The time history of the motion of a manipulator joint may be given in terms of what is known as a constraint-based trajectory. Let us now consider the analysis of such trajectories.

The positional constraints for a manipulator at two configurations can be satisfied by a trajectory of the form

$$\theta(t) = f(t)\theta_1 + [1 - f(t)]\theta_0 \qquad (11.28)$$

where

$$f(0) = 0 \qquad\qquad f(1) = 1$$

and θ_0 and θ_1 are the values of a joint variable at times $t = 0$ and $t = 1$, respectively. An important requirement is that the function $f(t)$ is continuous.

Let us define some specific polynomials that satisfy the conditions above. The simplest such function, $f(t) = t$, yields

$$\theta(t) = t\theta_1 + (1 - t)\theta_0 \qquad (11.29)$$

One drawback of this simple function is that the velocity is a constant and cannot be specified independently at the two boundaries of the trajectory. Infinite acceleration is also required between movements. Finally, the joint solution $\theta(t)$ may not always lie in the workspace.

The next trajectory to consider is a cubic having a symmetric form given by

$$\theta(t) = (1 - t)^2 \left[\theta_0 + \omega_0)t\right] + t^2 \theta_1 + \left[(2\theta_1 - \omega_1)(1 - t)\right] \qquad (11.30)$$

This cubic satisfies additional constraints on the initial and final velocities, namely

$$\dot{\theta}(0) = \omega_0 \qquad\qquad \dot{\theta}(1) = \omega_1$$

The cubic trajectory does not include a condition on the maximum attainable velocity. Furthermore, the acceleration cannot be specified independently.

A fifth-order polynomial enables the accelerations (α) at both ends of the trajectory to be specified in addition to the velocity and positional constraints. The equation is

$$\theta(t) = (1 - t)^3 \left[\theta_0 + (3\theta_0 + \omega_0)t + (\alpha_0 + 6\omega_0 + 12\theta_0)\frac{t^2}{2}\right]$$

$$+ t^3 \left[\theta_1 + (3\theta_1 - \omega_1)(1 - t) + (\alpha_1 - 6\omega_1 + 12\theta_1)\frac{(1 - t)^2}{2}\right]$$

$$(11.31)$$

A cosine trajectory satisfying the velocity constraints is given by

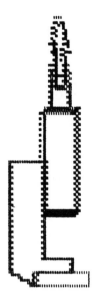

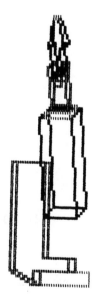

FIGURE 11.4   Simulation of a spherical robot joint motion.

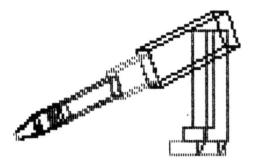

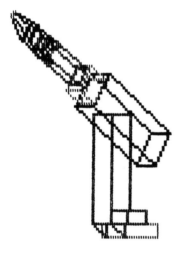

FIGURE 11.4   (continued)

$$\theta(t) = \cos^2\left[\frac{\pi}{2}(1 - t)\right]\left[\theta_1 - \frac{2\omega_1}{\pi} \cos \frac{\pi t}{2}\right]$$

$$+ \cos^2 \frac{\pi t}{2}\left\{\theta_0 + \frac{2\omega_0}{\pi} \cos\left[\frac{\pi}{2}(1 - t)\right]\right\}$$

(11.32)

A bang-bang trajectory with maximum acceleration and deceleration
is given by

$$\theta(t) = \begin{cases} \theta_0 + \dfrac{\alpha t^2}{2} & \text{for } 0 \leqslant t \leqslant \dfrac{1}{2} \\[3mm] \theta_1 - \dfrac{\alpha}{2}(1 - t^2) & \text{for } \dfrac{1}{2} \leqslant t \leqslant 1 \end{cases}$$

(11.33)

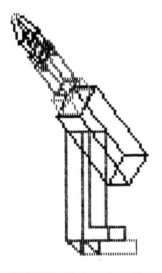

FIGURE 11.4   (continued)

TABLE 11.1 Programming Features of Painting Robots

Teaching method	Model	Type	Application	Special features
CP	RCP	705 855 856	Suitable for comparatively slow-speed painting, such as spray-gun painting of up to 500 mm/sec. Can paint either stationary or moving objects.	Up to 72 programs for each floppy disc. 24 min maximum working time per disc.
PTP(S)	RPS	705 855 856 846E 846F	Suitable for medium- to high-speed painting of stationary objects placed on a turntable. Programs can be modified partially with ease to correct mistakes. Available in two models: PTP(S) for equal-distance teaching and PTP(M) for unequal-distance teaching.	Up to 72 programs per floppy disc. Up to 11,808 memory points per floppy disc (72×164 points) in 5-axis type, and up to 9,792 points per floppy disc (72×136 points) in 6-axis type.
PTP(M)	RPM	705 855 856 846E 846F		Main memory is P-ROM and RAM (2,800 points for 5-axis type and 2,300 points for 6-axis type), and auxiliary memory is floppy discs. Program editing and copying function is equipped.
PTP(Z)	RPZ	705 855	Suitable for medium to high speed painting of objects	Up to 72 programs per floppy disc.

Model	Code	No.	Description	Features
		856, 846E, 846F	moving on a conveyor or the like. Programs can be readily modified in part to correct mistakes.	Up to 11,808 memory points per floppy disc (72 × 164 points) in 5-axis type, and up to 9792 points per floppy disc (72 × 136 points) in 6-axis type. Main memory is P-ROM and RAM (1300 points for 5-axis type and 1000 points for 6-axis type), and auxiliary memory is floppy discs. Program editing and copying functions are equipped.
CP + PTP(S)	RCS	705, 855, 856	Choice of either CP or PTP(S), or PTP(M), by simply turning a switch.	Both CP and PTP(S) features.
CP + PTP(M)	RCM	705, 855, 856		Both CP and PTP(M) features.
CP + PTP(Z)	RCZ	705, 855, 856	Choice of either CP or PTP(Z) by simply turning a switch.	Both CP and PTP(Z) features.
CP + PTP(M) + PTP(Z)	RPA	705, 855, 856	Has all the functions of CP, PTP(M) and PTP(Z), selected by simply turning a switch.	All the features of CP, PTP(M) and PTP(Z).

Source: Courtesy of Tokico America, Inc.

TABLE 11.2   Programming Control Methods for Selected Robots

Programming method	Robot	Control[a]	
		Nonservo	Servo
Mechanical setup	Mack Corporation	P	
	BASE Robot	P	
	Mobot Columnar	P	P
	Seiko Instruments Rimrock	P	
Pendant	Acco Jolly 8		P
	ASEA IR6-6		C, P
	Bendix SL 330	P	
	Binks		C, P
	Prab Robot F6		P
	Thermwood Series 3		P
Walk-through	DeVilbiss TR 3600S		C, P
	Nordson Robot System		C
	Planet Armax		C, P
	Prab Robot 4200	P	
	Thermwood Series 6		C

[a]P, Point-to-point; C, continuous path.

One of the advantages of constraint satisfaction trajectories is that only the initial and final conditions need be satisfied. A major difficulty is that the trajectory is weakly constrained, because the motion between the start and finish positions is not controlled. Interpolation schemes such as cubic splines enable obstacle avoidance in manipulator movements.

For any move of a manipulator, a typical trajectory characteristic consists of (1) an initial acceleration, (2) a constant velocity, and (3) a final deceleration (Fig. 11.4). This type of trajectory kinematics is commonly found in robots with computed path generation capabilities. In this trajectory plan, the computational capabilities of the control computer are used to define a path including

TABLE 11.3 Programming and Control Features of a Welding Robot

Model	IRBC-101
Motion control	CP by PTP teaching with linear and circular interpolations
Controlled axis	5 axis, simultaneously
Memory system	IC memory and cassette tape recorder for the external memory device
Memory capacity	About 800 teaching points at standard teaching conditions
Number of programs	Max. 100 programs stored in the memory
Welding speed control	Digital setting in value (cm/min)
Welding condition control	By digital keyboard setting
Welding conditions and setting ranges	Welding current, 30–35 A (2 A step) Arc voltage, 10–50 V (0.2 V step) Welding speed, 10–200 cm/min (1 cm/min step) [4–78.9 in./min (0.4 in./min step)] Preflow time and postflow time, 0, 0.5, 1.0, 1.5, 2.0, 2.5, 3.0 sec Dwell time and crater filling time, 0–9.9 sec (0.1 sec step)
Weaving control	Width, 0–20 mm (1 mm step), 0–0.8 in. (0.01 in. step) Frequency, max. 3 Hz (0.1 Hz step) Both end dwell time, 0–9.9 sec (0.1 sec step)
Input/output lines for jig control	10 pairs of input and output channels (expandable up to 100 pairs of channels—option)
Power sources	AC single-phase 1.5 kVA

Source: Courtesy of Merrick Engineering, Inc.

velocity and acceleration between taught points. If straight-line
paths are used, heavy objects may be smoothly transferred without
centrifugal and Coriolis forces being exerted. Because of the control on velocities and accelerations, high velocities may be used to
reduce the overall task cycle time. Control of both acceleration and
deceleration also improves the robot's service life by reducing jerky
motions and corresponding high forces. The programming features
of some industrial robots given in Tables 11.1 to 11.3 show the importance of trajectory considerations in robot utilization.

REVIEW QUESTIONS

1. Obtain the kinematic equations for the position of the hand of
   a number of industrial robots as assigned by the instructor,
   using direct trigonometric methods. Pick geometries that include
   revolute jointed, spherical, and cylindrical types. Study how
   the geometric complexities such as offsets in the shoulder and
   elbow joints affect the kinematic equations. By assuming a reasonable time history of the joint variables, determine the hand motion
   of the robot. Write a computer program to solve the hand coordinates given the joint motion of the robots assigned.
2. Derive the inverse kinematics solutions for the robots studied in
   problem 1. Obtain the time history of the joint coordinates for
   a given hand motion. Write a computer program for the inverse
   kinematics. Check your direct and inverse kinematics programs
   by using the output from one in the other. Comment on the
   multiple solutions obtained in the inverse kinematics.
3. Derive equations for the velocity components of the hand of a
   robot, given the joint motions. Develop a computer program for
   the velocity of the hand of a number of industrial robots.
4. Study a number of task-specific trajectories that are used in
   robots. Consider different applications such as painting, welding, gluing, and workcell operations.

FURTHER READING

Ahlberg, J. H., Nelson, E. N., and Walsh, J. L., *The Theory of
Splines and Their Application*, Academic Press, New York, 1967.

Ardayfio, D. D. and Kapur, R., "Microcomputer Aided Design Package for the Simulation of Industrial Robots," *Proc. 23rd Symp.
Mini and Microcomputers*, San Antonio, Texas, Dec. 1983.

Denavit, J. and Hartenberg, R. S., "A Kinematic Notation for
Lower Pair Mechanisms Based on Matrices," *J. Appl. Mech.*,
Vol. 22, *Trans. ASME*, Vol. 77, 1955, pp. 215–221.

Dombre, E., Barrel, P., and Liegeois, A., "A CAD System for Programming and Simulating Robots Actions," *Digital Syst. Ind. Autom.*, Vol. 2, No. 2, 1984, pp. 201–225.

Heginbotham, W. B., Bonner, M., and Kennedy, D. N., "Computer of Industrial Robots Performance by Interactive Computer Graphics," *9th Int. Symp. Industrial Robots*, Washington, DC, March 1979.

Heginbotham, W. B., Donner, M., and Kennedy, D. N., "Computer Graphics Simulation of Industrial Robot Interactions," 3rd CIRT, *7th Int. Symp. Industrial Robots*, 1977.

Ho, C. Y. and Cook, C. C., "The Application of Spline Functions to Trajectory Generation for Computer Controlled Manipulators," Interdepartmental Memo, Department of Computer Science, University of Missouri-Rolla, 1980.

Horn, B. K. P. and Inoue, H., "Kinematics of the MIT-AI-VICARM Manipulator," MIT Artificial Intelligency Laboratory, Working Paper 69, May 1974.

Kretch, S. J., "CAD/CAM for Robots," Robots VI Conference, Society of Manufacturing Engineers, Detroit, 1982.

Lewis, R. A., "Autonomous Manipulation of a Robot: Summary of Manipulator Software Functions," Jet Propulsion Laboratory, NASA Tech. Memo. 33-679, March 1974.

Liegeois, A., Fournier, A., Aldon, N. J., and Borrel, P., "A System for Computer-Aided Design of Robots and Manipulators," *10th Int. Symp. Industrial Robots*, Milan, March 1980, pp. 441–452.

Menz, G., "PUMA Simulation Program," Technical Report, Center for Manufacturing Productivity, Rensselaer Polytechnic Institute, 1981.

Mujtaba, M. S., "Discussion of Trajectory Calculation Methods," Exploratory Study of Computer Integrated Assembly Systems, Stanford Artificial Intelligence Laboratory Progress Report, June 1977.

Norton, R. L., "Graphic Simulation of PUMA Robot Motions on the Apple Computer," *Proc. 3rd Int. Conf. Computers in Engineering*, Chicago, Vol. 2, Aug. 1983, pp. 126–130.

Paul, R., "Manipulator Path Control," *Proc. 1975 Int. Conf. Cybernetics Society*, Sept. 1975.

Paul, R., "The Mathematics of Computer Controlled Manipulators," *Proc. 1977 Joint Automatic Control Conf.*, 1977.

Paul, R. C., Shimano, B., and Mayer, G. E., "Kinematic Controlled Equations for Simple Manipulators," *IEEE Trans. Syst. Man. Cybern.*, Vol. SMC-11, No. 6, June 1981.

Pieper, D. L., "The Kinematics of Manipulators under Computer Control." Ph.D. Dissertation, Stanford University, Department of Mechanical Engineering, Oct. 1968.

Pieper, D. L. and Roth, B., "The Kinematics of Manipulators under Computer Control." *Proc. 2nd Int. Cong. Theory of Machines and Mechanisms*, Vol. 2, 1969, pp. 159-168.

Roth, B., "Performance Evaluation of Manipulators from a Kinematic Viewpoint," NBS Special Publications: *Performance Evaluation of Programmable Robots and Manipulators* (Report of Workshop held Oct. 1975), National Bureau of Standards, Washington, DC.

Roth, B., Rastegar, J., and Scheinman, V., *On the Theory and Practice of Robots and Manipulators*, Vol. 1, Springer-Verlag, New York, 1974, pp. 93-113.

12
Dynamics and Control

The dynamics of a robot can be formulated conveniently by two techniques. The first is the Lagrange's method, which uses the kinetic and potential energy expressions in terms of the generalized coordinates. The second is the well-known Newton's approach for dynamic systems. In this section we will determine the joint torques and forces for a robot by these two methods. The Lagrangian method will be applied to revolute-jointed two-link and three-link robots. The expressions for the kinetic energy and potential energy are derived by a direct method. The resulting dynamic equations are then given in terms of the dynamic coefficients. Other robot geometries containing prismatic joints are then considered. We finally present the detailed steps in generating the dynamic equations by using the Newton-Euler formulation for a typical two-link robot.

12.1 LAGRANGIAN METHOD FOR A TWO-LINK ROBOT

Consider the two-link robot shown in Figure 12.1. The first step in the Lagrangian method is to obtain the expressions for the kinetic and potential energy.

For mass m_1 the kinetic and potential energy expressions are

$$K_1 = \frac{1}{2}m_1 l_1^2 \dot{\theta}_1^2$$

$$P_1 = +m_1 g l_1 \cos \theta_1 \qquad (12.1)$$

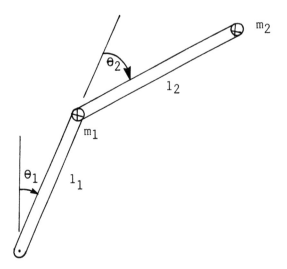

FIGURE 12.1   Two-link robot with lumped masses m_1 and m_2.

The coordinates of mass m_2 are given by

$$x_2 = l_1 \sin \theta_1 + l_2 \sin(\theta_1 + \theta_2)$$

$$y_2 = l_1 \cos \theta_1 + l_2 \cos(\theta_1 + \theta_2) \qquad (12.2)$$

The velocity components are

$$\dot{x}_2 = l_1 \dot{\theta}_1 \cos \theta_1 + l_2(\dot{\theta}_1 + \dot{\theta}_2) \cos(\theta_1 + \theta_2)$$

$$\dot{y}_2 = -l_1 \dot{\theta}_1 \sin \theta_1 - l_2(\dot{\theta}_1 + \dot{\theta}_2) \sin(\theta_1 + \theta_2) \qquad (12.3)$$

so that

$$v_2^2 = \dot{x}_2^2 + \dot{y}_2^2$$

$$= l_1 \dot{\theta}_1^2 + l_2^2(\dot{\theta}_1 + \dot{\theta}_2)^2 + 2l_1 l_2 \dot{\theta}_1(\dot{\theta}_1 + \dot{\theta}_2) \cos \theta_2 \qquad (12.4)$$

The expression for the kinetic energy is thus

$$K_2 = \frac{1}{2}m l_1^2 \dot{\theta}_1^2 + \frac{1}{2}m_2 l_2^2(\dot{\theta}_1 + \dot{\theta}_1)^2 + m_2 l_1 l_2 \dot{\theta}_1(\dot{\theta}_1 + \dot{\theta}_2) \cos \theta_2$$

$$(12.5)$$

The potential energy is

$$P_2 = +m_2 g l_1 \cos \theta_1 + m_2 g l_2 \cos(\theta_1 + \theta_2) \tag{12.6}$$

Dynamic Equations

The derivatives needed in the Lagrangian formulation can be obtained by writing $K = K_1 + K_2$ and $P = P_1 + P_2$. Then

$$\frac{\partial K}{\partial \dot\theta_1} = (m_1 + m_2)l_1^2 \dot\theta_1 + m_2 l_2^2 \dot\theta_1 + m_2 l_2^2 \dot\theta_2 \tag{12.7}$$

$$+ 2m_2 l_1 l_2 \dot\theta_1 \cos \theta_2 + m_2 l_1 l_2 \dot\theta_2 \cos \theta_2$$

$$\frac{d}{dt}\left(\frac{\partial K}{\partial \dot\theta_1}\right) = [(m_1 + m_2)l_1^2 + m_2 l_2^2 + 2m_2 l_1 l_2 \cos \theta_2]\ddot\theta_1$$

$$+ (m_2 l_2^2 + m_2 l_1 l_2 \cos \theta_2)\ddot\theta_2$$

$$-2m_2 l_1 l_2 \dot\theta_1 \dot\theta_2 \sin \theta_2 - m_2 l_1 l_2 \dot\theta_2^2 \sin \theta_2$$

$$\frac{\partial P}{\partial \theta_1} = -m_1 g l_1 \sin \theta_1 - m_2 g l_1 \sin \theta_1 - m_2 g l_2 \sin(\theta_1 + \theta_2)$$

The torque at joint 1 is thus obtained from the Lagrangian equation

$$T_1 = \frac{d}{dt}\frac{\partial K}{\partial \dot\theta_1} - \frac{\partial K}{\partial \theta_1} + \frac{\partial P}{\partial \theta_1} \tag{12.8}$$

as

$$T_1 = [(m_1 + m_2)l_1^2 + m_2 l_2^2 + 2m_2 l_1 l_2 \cos \theta_2]\ddot\theta_1$$

$$+ (m_2 l_2^2 + m_2 l_1 l_2 \cos \theta_2)\ddot\theta_2$$

$$-2m_1 l_1 l_2 \dot\theta_1 \dot\theta_2 \sin \theta_2 - m_2 l_1 l_2 \dot\theta_2^2 \sin \theta_2$$

$$- (m_1 + m_2)g l_1 \sin \theta_1 - m_2 g l_2 \sin(\theta_1 + \theta_2) \tag{12.9}$$

The torque at joint 2 is obtained similarly from the following terms:

$$\frac{\partial K}{\partial \dot{\theta}_2} = m_2 l_1^2 \dot{\theta}_1 + m_2 l_2^2 (\dot{\theta}_1 + \dot{\theta}_2) + m_2 l_1 l_2 (2\dot{\theta}_1 + \dot{\theta}_2) \cos \theta_2$$

$$\frac{d}{dt}\left(\frac{\partial K}{\partial \dot{\theta}_2}\right) = m_2 (l_1^2 + l_2^2 + 2 l_1 l_2 \cos \theta_2) \ddot{\theta}_1 + m_2 l_2^2 \ddot{\theta}_2$$

$$- m_2 l_1 l_2 (2\dot{\theta}_1 \dot{\theta}_2 + \dot{\theta}_2^2) \sin \theta_2$$

$$\frac{\partial K}{\partial \theta_2} = -m_2 l_1 l_2 \dot{\theta}_1 (\dot{\theta}_1 + \dot{\theta}_2) \sin \theta_2$$

$$\frac{\partial P}{\partial \theta_2} = -m_2 g l_2 \sin(\theta_1 + \theta_2)$$

Finally,

$$T_2 = m_2 (l_2^2 + l_1 l_2 \cos \theta_2) \ddot{\theta} + m_2 l_2^2 \ddot{\theta}_2$$
$$+ m_2 l_1 l_2 \dot{\theta}_2^2 \sin \theta_2 - m_2 g l_2 \sin(\theta_1 + \theta_2)$$

$$(12.10)$$

Dynamic Coefficients

The equations for the joint torques can be written as

$$T_1 = D_{11}\ddot{\theta}_1 + D_{12}\ddot{\theta}_2 + D_{111}\dot{\theta}_1^2 + D_{122}\dot{\theta}_2^2$$
$$+ D_{112}\dot{\theta}_1\dot{\theta}_2 + D_{121}\dot{\theta}_2\dot{\theta}_1 + D_1$$

$$T_2 = D_{21}\ddot{\theta}_1 + D_{22}\ddot{\theta}_2 + D_{211}\dot{\theta}_1^2 + D_{222}\dot{\theta}_2^2 + D_{212}\dot{\theta}_1\dot{\theta}_2$$
$$+ D_{221}\dot{\theta}_2\dot{\theta}_1 + D_2 \qquad\qquad (12.11)$$

The coefficients D_i, D_{ij}, and D_{ijk} are known as the dynamic coefficients and are defined in the following way:

$D_{ii}\ddot{\theta}_i$ is the effective inertia at joint i.

$D_{ij}\dot{\theta}_i\dot{\theta}_j$ is the coupling inertia between joints i and j.

$D_{ijj}\dot{\theta}_j^2$ is the centripetal force acting at joint i due to a velocity at joint j.

$D_{ijk}\dot{\theta}_j\dot{\theta}_k$ is the Coriolis force acting at joint i due to velocities at j and k.

D_i is the gravity force.

The dynamic coefficients for the two-link robot can therefore be obtained from the dynamic equations as

$$D_{11} = (m_1 + m_2)l_1^2 + m_2 l_2^2 + 2m_2 l_1 l_2 \cos \theta_2$$

$$D_{22} = m_2 l_2 (l_2 + l_1)\cos \theta_2 \qquad D_{211} = m_2 l_1 l_2 \sin \theta_2$$

$$D_{12} = m_2 l_2^2 + m_2 l_1 l_2 \cos \theta_2 \qquad D_{222} = 0$$

$$D_{122} = -m_2 l_1 l_2 \sin \theta_2 \qquad D_{112} = D_{121} = -m_2 l_1 l_2 \sin \theta_2$$

$$D_{111} = 0 \qquad\qquad D_{212} = D_{221} = 0$$

$$D_1 = -(m_1 + m_2)gl_1 \sin \theta_1 - m_2 gl_2 \sin(\theta_1 + \theta_2)$$

$$D_2 = -m_2 l_2 g \sin(\theta_1 + \theta_2) \tag{12.12}$$

In configuration space form these dynamic equations are

$$[A]\begin{bmatrix}\ddot{\theta}_1 \\ \ddot{\theta}_2\end{bmatrix} + [B]\begin{bmatrix}\dot{\theta}_1^2 \\ \dot{\theta}_2^2\end{bmatrix} + [C]\begin{bmatrix}\dot{\theta}_1\dot{\theta}_2 \\ \dot{\theta}_2\dot{\theta}_1\end{bmatrix} + \begin{bmatrix}G_1 \\ G_2\end{bmatrix} = \begin{bmatrix}T_1 \\ T_2\end{bmatrix} \tag{12.13}$$

where

$$[A] = \begin{bmatrix} (m_1 + m_2)l_1^2 + m_2 l_2^2 + 2m_2 l_1 l_2 \cos \theta_2 & m_2 l_2^2 + m_2 l_1 l_2 \cos \theta_2 \\ m_2(l_1^2 + l_2^2 + 2l_1 l_2 \cos \theta_2) & m_2 l_2^2 \end{bmatrix}$$

$$[B] = \begin{bmatrix} 0 & -m_2 l_1 l_2 \sin \theta_2 \\ m_2 l_1 l_2 \sin \theta_2 & 0 \end{bmatrix}$$

$$[C] = \begin{bmatrix} -m_2 l_1 l_2 \sin \theta_2 & -m_2 l_1 l_2 \sin \theta_2 \\ 0 & 0 \end{bmatrix}$$

$$[G] = \begin{bmatrix} -(m_1 + m_2) g l_1 \sin \theta_1 - m_2 g l_2 \sin(\theta_1 + \theta_2) \\ -m_2 g l_2 \sin(\theta_1 + \theta_2) \end{bmatrix}$$

12.2 LAGRANGIAN METHOD FOR A THREE-LINK ROBOT

The coordinates of M_2 are (Figure 12.2)

$$x_2 = l_2 C_1 S_2$$

$$y_2 = l_2 S_1 S_2$$

$$z_2 = l_2 C_2 \tag{12.14}$$

where $S_i = \sin \theta_i$ and $C_i = \cos \theta_i$

The velocity components are

$$\dot{x}_2 = l_2(-S_1 \dot{\theta}_1 S_2 + C_1 C_2 \dot{\theta}_2)$$

$$\dot{y}_2 = l_2(C_1 \dot{\theta}_1 S_2 + S_1 C_2 \dot{\theta}_2)$$

$$\dot{z}_2 = l_2(-S_2 \dot{\theta}_2) \tag{12.15}$$

so that

$$v^2 = \dot{x}_2^2 + \dot{y}_2^2 + \dot{z}_2^2$$

$$= l_2^2[(-S_1 \dot{\theta}_1 S_2 + C_1 C_2 \dot{\theta}_2)^2 + (C_1 \dot{\theta}_1 S_2 + S_1 C_2 \dot{\theta}_2)^2 + (S_2 \dot{\theta}_2)^2]$$

$$= l_2^2(s_2^2 \dot{\theta}_1^2 + \dot{\theta}_2^2) \tag{12.16}$$

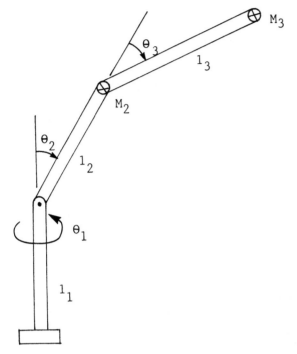

FIGURE 12.2   Three-link robot configuration with concentrated masses.

From these, the kinetic energy of M_2 is thus

$$K_2 = \frac{1}{2}M_2 l_2^{\ 2}(S_2^{\ 2}\dot{\theta}_1^{\ 2} + \dot{\theta}_2^{\ 2}) \qquad (12.17)$$

The potential energy is

$$P_2 = M_2 g l_2 C_2 \qquad (12.18)$$

The coordinates of M_3 are

$$x_3 = C_1(l_2 S_2 + l_3 S_{23})$$

$$y_3 = S_1(l_2 S_2 + l_3 S_{23})$$

$$z_3 = l_2 C_2 + l_3 C_{23} \qquad (12.19)$$

The velocity components are

$$\dot{x}_3 = -S_1\dot{\theta}_1(l_2S_2 + l_3S_{23}) + C_1[l_2C_2\dot{\theta}_2 + l_3C_{23}(\dot{\theta}_2 + \dot{\theta}_3)]$$

$$\dot{y}_3 = C_1\dot{\theta}_1(l_2S_2 + l_3S_{23}) + S_1[l_2C_2\dot{\theta}_2 + l_3C_{23}(\dot{\theta}_2 + \dot{\theta}_3)]$$

$$\dot{z}_3 = -l_2S_2\dot{\theta}_2 - l_3S_{23}(\dot{\theta}_2 + \dot{\theta}_3) \tag{12.20}$$

so that

$$v_3^2 = \dot{x}_3^2 + \dot{y}_3^2 + \dot{z}_3^2$$

$$= \dot{\theta}_1^2(l_2S_2 + l_3S_{23})^2 + [l_2C_2\dot{\theta}_2 + l_3(C_{23}(\dot{\theta}_2 + \dot{\theta}_2 + \dot{\theta}_3)]^2$$

$$+[-l_2S_2\dot{\theta}_2 - l_3S_{23}(\dot{\theta}_2 + \dot{\theta}_3)]^2 \tag{12.21}$$

$$= (l_2S_2 + l_3S_{23})^2\dot{\theta}_1^2 + l_2^2\dot{\theta}_2^2 + 2l_2l_3C_3\dot{\theta}_2(\dot{\theta}_2 + \dot{\theta}_3) + l_3^2(\dot{\theta}_2 + \dot{\theta}_3)^2$$

The kinetic energy of M_3 is thus

$$K_3 = \frac{1}{2}M_3(l_2S_2 + l_3S_{23})^2\dot{\theta}_1^2 + \frac{1}{2}M_3(l_2^2 + 2l_2l_3C_3 + l_3^2)\dot{\theta}_2^2$$

$$+ M_3l_3(l_2C_3 + l_3)\dot{\theta}_2\dot{\theta}_3 + \frac{1}{2}M_3l_3^2\dot{\theta}_3^2 \tag{12.22}$$

The potential energy is

$$P_3 = M_3g(l_2C_2 + l_3C_{23}) \tag{12.23}$$

The Lagrangian function is

$$L = \frac{1}{2}M_2l_2^2(S_2^2\dot{\theta}_1^2 + \dot{\theta}_2^2) + \frac{1}{2}M_3(l_2S_2 + l_3S_{23})^2\dot{\theta}_1^2$$

$$+ \frac{1}{2}M_3(l_2^2 + 2l_2l_3C_3 + l_3^2)\dot{\theta}_2^2 + M_3l_3(l_2C_3 + l_3)\dot{\theta}_2\dot{\theta}_3$$

$$+ \frac{1}{2}M_3l_3^2\dot{\theta}_3^2 - M_2gl_2C_2 - M_3g(l_2C_2 + l_3C_{23}) \tag{12.24}$$

The expressions needed in the Lagrangian formulation are

$$\frac{\partial L}{\partial \dot\theta_1} = M_2 l_2^2 S_2^2 \dot\theta_1 + M_3 (l_2 S_2 + l_3 S_{23})^2 \dot\theta_1$$

$$\frac{d}{dt}\left(\frac{\partial L}{\partial \dot\theta_2}\right) = M_2 l_2^2 S_2^2 \ddot\theta_1 + M_3 (l_2 S_2 + l_3 S_{23})^2 \ddot\theta_1$$

$$+ 2M_2 l_2^2 S_2 C_2 \dot\theta_1 \dot\theta_2 + 2M_3 (l_2 S_2 + l_3 S_{23})$$

$$[l_2 C_2 \dot\theta_2 + l_3 C_{23}(\dot\theta_2 + \dot\theta_3)]\dot\theta_1$$

$$\frac{\partial L}{\partial \theta_1} = 0$$

$$\frac{\partial L}{\partial \dot\theta_2} = M_2 l_2^2 \dot\theta_2 + M_3 (l_2^2 + 2l_2 l_3 C_3 + l_3^2) \dot\theta_2 + M_3 l_3 (l_2 C_3 + l_3) \dot\theta_3$$

$$\frac{d}{dt}\left(\frac{\partial L}{\partial \dot\theta_2}\right) = M_2 l_2^2 \ddot\theta_2 + M_3 (l_2 + 2l_2 l_3 C_3 + C_3) \ddot\theta_2 + M_3 l_3 (l_2 C_3 + l_3) \ddot\theta_3$$

$$-2M_3 l_2 l_3 S_3 \dot\theta_3 \dot\theta_2 - M_3 l_3 l_2 S_3 \dot\theta_3^2$$

$$\frac{\partial L}{\partial \theta_2} = M_2 l_2^2 S_2 C_2 \dot\theta_1^2 + M_3 (l_2 S_2 + l_3 S_{23})(l_2 C_2 + l_3 C_{23}) \dot\theta_1^2$$

$$+ M_2 g l_2 S_2 + M_3 g (l_2 S_2 + l_3 S_{23})$$

$$\frac{\partial L}{\partial \dot\theta_3} = M_3 l_3 (l_2 C_3 + l_3) \dot\theta_2 + M_3 l_3^2 \dot\theta_3$$

$$\frac{d}{dt}\left(\frac{\partial L}{\partial \dot\theta_3}\right) = M_3 l_3 (l_2 C_3 + l_3) \ddot\theta_2 + M_3 l_3^2 \ddot\theta_2 + M_3 l_3 (-l_2 S_3) \dot\theta_3 \dot\theta_2$$

$$\frac{\partial L}{\partial \theta_3} = M_3 (l_2 S_2 + l_3 S_{23}) l_3 C_{23} \dot\theta_1^2 - M_3 l_2 l_3 S_3 \dot\theta_2^2$$

$$- M_3 l_3 l_2 S_3 \dot\theta_2 \dot\theta_3 + M_3 g (l_3 S_{23}) \tag{12.25}$$

The equations for the joint torques are obtained from these expressions as

$$T_1 = [M_2 l_2^2 S_2^2 + M_3 (l_2 S_2 + l_3 S_{23})^2] \ddot{\theta}_1$$

$$+ 2[M_2 l_2^2 S_2 C_2 + M_3 (l_2 S_2 + l_3 S_{23})(l_2 C_2 + l_3 C_{23})] \dot{\theta}_1 \dot{\theta}_2$$

$$+ 2M_3 l_3 C_{23} (l_2 S_2 + l_3 S_{23}) \dot{\theta}_3 \dot{\theta}_1$$

$$T_2 = [M_2 l_2^2 + M_3 (l_2^2 + 2l_2 l_3 C_3 + l_3^2)] \ddot{\theta}_2$$

$$+ M_3 l_3 (l_2 C_3 + l_3) \ddot{\theta}_3 - 2M_3 l_2 l_3 S_3 \dot{\theta}_3 \dot{\theta}_2 - M_3 l_2 l_3 S_3 \dot{\theta}_3^2$$

$$- [M_2 l_2^2 S_2 C_2 + M_3 (l_2 S_2 + l_3 S_{23})(l_2 C_2 + l_3 C_{23})] \dot{\theta}_1^2$$

$$- M_2 g l_2 S_2 - M_3 g (l_2 S_2 + l_3 S_{23})$$

$$T_3 = M_3 l_3 (l_2 C_3 + l_3) \ddot{\theta}_2 + M_3 l_3^2 \ddot{\theta}_3 - M_3 l_3 \dot{\theta}_2 \dot{\theta}_3$$

$$- M_3 (l_2 S_2 + l_3 S_{23}) l_3 C_{23} \dot{\theta}_1^2$$

$$+ M_3 l_2 l_3 S_3 \dot{\theta}_2^2 + M_3 l_3 \dot{\theta}_2 \dot{\theta}_3 - M_3 g l_3 S_{23} \qquad (12.26)$$

The dynamic coefficients are

$$D_{11} = M_2 l_2^2 S_2^2 + M_3 (l_2 S_2 + l_3 S_{23})^2$$

$$D_{12} = D_{13} = D_1 = 0$$

$$D_{112} = D_{121} = M_2 l_2^2 S_2 C_2 + M_3 (l_2 S_2 + l_3 S_{23})(l_2 C_2 + l_3 C_{23})$$

$$D_{113} = D_{131} = M_3 (l_2 S_2 + l_3 S_{23}) l_3 C_{23}$$

$$D_{21} = 0$$

$$D_{22} = M_2 l_2^2 + M_3 (l_2^2 + 2l_2 l_3 C_3 + l_3^2)$$

$$D_{23} = M_3 l_3 (l_2 C_3 + l_3)$$

$$D_{212} = D_{221} = D_{213} = D_{231} = D_{222} = 0$$

$$D_{211} = -M_2 l_2{}^2 S_2 C_2$$

$$D_{223} = D_{232} = -2M_2 l_2 l_3 S_3$$

$$D_{233} = -M_3 l_2 l_3 S_3$$

$$D_2 = -M_2 g l_2 S_2 - M_3 g (l_2 S_2 + l_3 S_{23})$$

$$D_{31} = 0$$

$$D_{32} = M_3 l_3 (l_2 C_3 + l_3)$$

$$D_{33} = M_3 l_3{}^2$$

$$D_{322} = -M_3 (l_2 S_2 + l_3 S_{23}) l_3 C_{23}$$

$$D_{322} = M_3 l_2 l_3 S_3$$

$$D_{323} = D_{332} = D_{333} = D_{312} = D_{321} = 0$$

$$D_3 = -M_3 g (l_3 S_{23}) \tag{12.27}$$

12.3 DISTRIBUTED MASS EFFECTS

The upright column may be modeled as a uniform cylinder with height h and radius r_1. The kinetic energy of the upright column is

$$K_1 = \frac{1}{2} \left(\frac{m_1 r_1{}^2}{2} \right) \dot{\theta}_1{}^2 \tag{12.30}$$

where the term $m_1 r_1{}^2 / 2$ is the moment of inertia. The second link is modeled as a thin rod of mass m_2 and length l_2. The velocity of an infinitesimal mass located at a distance σ on link 2 is

$$V = \sigma[(-S_1 S_2 \dot\theta_1 + C_1 C_2 \dot\theta_2)\hat{i} + (C_1 S_2 \dot\theta_1 + S_1 C_2 \dot\theta_2)\hat{j} + (-S_2 \dot\theta_2)\hat{k}]$$

from which

$$v^2 = \sigma^2(S_2^2 \dot\theta_1^2 + \dot\theta_2^2) \tag{12.31}$$

The kinetic energy of link 2 is thus

$$K_2 = \frac{1}{2}\frac{M_2}{l_2}(S_2^2 \dot\theta_1^2 + \dot\theta_2^2)\int_0^{l_2} \sigma^2 \, d\sigma$$

$$= \frac{1}{6} M_2 l_2^2 (S_2^2 \dot\theta_1^2 + \dot\theta_2^2) \tag{12.32}$$

For an infinitesimal particle located by σ on link 3

$$v^2 = (l_2 S_2 + \sigma S_{23})^2 \dot\theta_1^2 + [l_2 C_2 \dot\theta_2 + \sigma C_{23}(\dot\theta_2 + \dot\theta_3)]^2$$

$$+ [+l_2 S_2 \dot\theta_2 + \sigma S_{23}(\dot\theta_2 + \dot\theta_3)]^2$$

$$= (l_2 S_2^2 + 2l_2 S_2 S_{23}\sigma + S_{23}^2 \sigma^2)\dot\theta_1^2$$

$$+ l_2^2 \dot\theta_2^2 + 2l_2 C_3 \dot\theta_2(\dot\theta_2 + \dot\theta_3)\sigma + (\dot\theta_2 + \dot\theta_3)^2 \sigma^2 \tag{12.33}$$

The kinetic energy of link 3 follows as

$$K_3 = \frac{1}{2}\frac{M_3}{l_3}\int_0^{l_3} v^2 \, d\sigma$$

$$= \frac{M_3}{2}\left(l_2 S_2^2 + l_2 l_3 S_2 S_{23} + \frac{l_3^2}{3} S_{23}^2\right)\dot\theta_1^2$$

$$\frac{M_3}{2}\left(l_2^2 + l_2 l_3 C_3 + \frac{l_3^2}{3} \dot\theta_2^2\right)$$

$$+ \frac{M_3}{2}\left(l_2 l_3 C_3 + \frac{l_3^2}{3}\right)\dot\theta_2 \dot\theta_3 + \frac{M_3 l_3^2}{6}\dot\theta_3^2 \tag{12.34}$$

The potential energy is

$$P = \frac{m_2 g l_2}{2} C_2 + m_3 g \left(l_2 C_2 + \frac{l_3}{2} C_{23} \right) \qquad (12.35)$$

The final equations for the torques are

$$T_1 = I_1 \ddot{\theta}_1 + \frac{1}{3} m_2 l_2^2 (S_2^2 \ddot{\theta}_1 + 2 S_2 C_2 \dot{\theta}_1 \dot{\theta}_2)$$

$$+ m_3 \left(l_2^2 S_2^2 + l_2 l_3 S_2 S_{23} + \frac{l_3^2}{3} S_{23}^2 \right) \ddot{\theta}_1$$

$$+ m_3 \left[2 l_2^2 S_2 C_2 + l_2 l_3 (C_2 S_{23} + S_2 C_{23}) + \frac{2 l_3^2}{3} S_{23} C_{23} \right] \dot{\theta}_1 \dot{\theta}_2$$

$$+ m_3 \left(l_2 l_3 S_2 C_{23} + \frac{2 l_3^2}{3} S_{23} C_{23} \right) \dot{\theta}_3 \dot{\theta}_1 \qquad (12.36)$$

$$T_2 = \frac{1}{3} m_2 l_2^2 (\ddot{\theta}_2 - S_2 C_2 \dot{\theta}_1^2) + m_3 \left(l_2^2 + l_2 l_3 C_3 + \frac{l_3^2}{3} \right) \ddot{\theta}_2$$

$$+ \frac{m_3}{2} \left(l_2 l_3 C_3 + \frac{2 l_3^2}{3} \right) \ddot{\theta}_3 - \frac{m_3}{2} \left[2 l_2^2 S_2 C_2 + l_2 l_3 (C_2 S_{23} + S_2 C_{23}) \right.$$

$$\left. + \frac{2 l_3^2}{3} S_{23} C_{23} \right] \dot{\theta}_1^2$$

$$+ 2 m_3 l_2 l_3 S_3 \dot{\theta}_2 \dot{\theta}_3 - m_3 l_2 l_3 S_3 \dot{\theta}_3^2 + m_2 g \frac{l_2}{2} S_2 +$$

$$+ m_3 g \left(l_2 S_2 + \frac{l_3}{2} S_{23} \right) \qquad (12.37)$$

$$T_3 = + \frac{m_3}{2} \left(l_2 l_3 C_3 + \frac{2 l_3^2}{3} \right) \ddot{\theta}_2 + \frac{m_3 l_3^2}{3} \ddot{\theta}_3$$

$$- \frac{m_3}{2}\left(l_2 l_3 S_2 C_{23} + \frac{2l_3^2}{3} S_{23} C_{23}\right)\dot{\theta}_1^2 + \frac{m_3}{2}l_2 l_3 S_3 \dot{\theta}_2^2$$

$$+ \frac{m_3 g l_3}{2} S_{23}$$

12.4 LAGRANGIAN RESULTS FOR OTHER ROBOT TYPES

Spherical Robot

For a spherical manipulator the Lagrangian function can be derived as

$$L = \frac{1}{2}m_2(s_2^2\dot{\theta}_1^2 + \dot{\theta}_2^2)l_2^2 + \frac{1}{2}m_3(\dot{d}_3^2 + d_3^2\dot{\theta}_2^2 + s_2^2 d_3^2\dot{\theta}_1^2) - m_2 g l_2 c_2$$

$$-m_3 g d_3 c_2 \qquad\qquad (12.39)$$

The equations of motion are

$$T_1 = (m_2 l_2^2 + m_3 d_3^2)s_2^2\ddot{\theta}_1 + 2(m_2 l_2^2 + m_3 d_3^2)s_2 c_2 \dot{\theta}_1 \dot{\theta}_2$$

$$2m_3 s_2^2 d_3 \dot{\theta}_1 \dot{d}_3 \qquad\qquad (12.40)$$

$$T_2 = (m_2 l_2^2 + m_3 d_3^2)\ddot{\theta}_2 + 2m_3 d_3 \dot{\theta}_2 \dot{d}_3 - (m_2 l_2^2 + m_3 d_3^2)s_2 c_2 \dot{\theta}_1^2$$

$$- (m_2 l_2 + m_3 d_3)g s_2 \qquad\qquad (12.41)$$

$$F_3 = m_3 \ddot{d}_3 - m_3 d_3 \dot{\theta}_2^2 - m_3 s_2^2 d_3 \dot{\theta}_1^2 + m_3 g c_2 \qquad\qquad (12.42)$$

The corresponding nonzero dynamic coefficients are

$$D_{11} = (m_2 l_2^2 + m_3 d_3^2)s_2^2$$

$$D_{22} = (m_2 l_2^2 + m_3 d_3^2) \qquad D_2 = -(m_2 l_2 + m_3 d_3)g s_2$$

$$D_{33} = m_3 \qquad D_3 = m_3 g c_2$$

$$D_{112} = D_{121} = (m_2 l_2^2 + m_3 d_3) s_2$$

$$D_{113} = D_{131} = m_3 s_2^2 d_3$$

$$D_{211} = -(m_2 l_2^2 + m_3 d_3^2) s_2 c_2$$

$$D_{311} = -m_3 d_3 s_2^2 \qquad D_{322} = -m_3 d_3 \qquad (12.43)$$

Cylindrical Robot

Considering a mass m_2 at joint 3 and a payload m_3 at the end of link 3, the Lagrangian can be shown to be

$$L = \frac{1}{2} m_2 \dot{d}_2^2 + \frac{1}{2} m_3 (\dot{d}_3^2 + d_3^2 \dot{\theta}_1^2 + \dot{d}_2^2) - m_2 g d_2 - m_3 g_2 d_2 \quad (12.44)$$

The equations of motion are thus

$$T_1 = m_3 (2 d_3 \dot{d}_3 \dot{\theta}_1 + d_3^2 \ddot{\theta}_1)$$

$$F_2 = (m_2 + m_3) \ddot{d}_2 + (m_2 + m_3) g$$

$$F_3 = m_3 \ddot{d}_3 - m_3 d_3 \dot{\theta}_1^2 \qquad (12.45)$$

From these equations the dynamic coefficients are obtained as

$$D_{11} = m_3 d_3^2 \qquad D_{131} = m_3 d_3 \qquad D_{113} = m_3 d_3$$

$$D_{22} = m_2 + m_3 \qquad D_2 = -(m_2 + m_3) g$$

$$D_{33} = m_3 \qquad D_{311} = -m_3 d_3 \qquad (12.46)$$

The dynamic equations derived in this section can be used with an appropriate trajectory scheme for the joint motions, as disucssed in the preceding chapter. Figures 12.3 to 12.5 show the mass distribution needed in the dynamic models of three goemetric types of robots. A sample plot of the joint torques or forces using the computed path trajectory shown in Figure 12.6 is given in Figures 12.7 and 12.8.

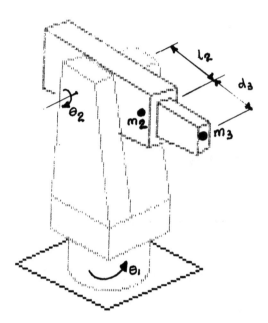

FIGURE 12.3   Spherical robot.

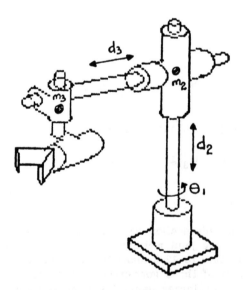

FIGURE 12.4   Cylindrical robot.

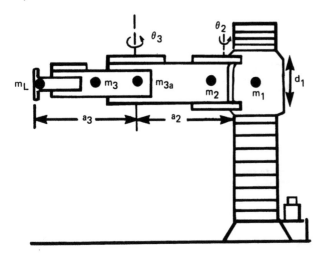

FIGURE 12.5   Dynamic model of a prosthetic robot.

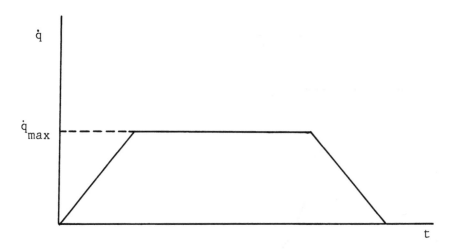

FIGURE 12.6   Velocity-time plot of a computed path trajectory.

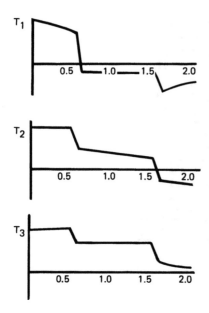

FIGURE 12.7   Torque curves for an anthropomorphic robot.

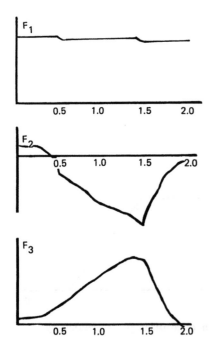

FIGURE 12.8   Torque curves for a cylindrical robot.

12.5 NEWTON-EULER FORMULATION

For the two-link robot shown in Figure 12.9, the kinematic quantities
needed in the Newton-Euler dynamic formulation are (vector notation)

$$r_1 = \frac{1}{2}l_1(\cos \phi_1 \hat{i} + \sin \phi_1 \hat{j})$$

$$r_2 = 2r_1 + \frac{1}{2}l_2[\cos(\phi_1 + \phi_2) \hat{i} + \sin(\phi_1 + \phi_2)\hat{j}]$$

$$\dot{r}_1 = \frac{1}{2}l_1\dot{\phi}_1(-\sin \phi_1 \hat{i} + \cos \phi_1 \hat{j})$$

$$\dot{r}_2 = 2\dot{r}_1 + \frac{1}{2}l_2(\dot{\phi}_1 + \dot{\phi}_2)[-\sin(\phi_1 + \phi_2) \hat{i} + \cos(\phi_1 + \phi_2) \hat{j}]$$

$$\ddot{r}_1 = \frac{1}{2}l_1[-(\ddot{\phi}_1 \sin \phi_1 + \dot{\phi}_1^2 \cos \phi_1) \hat{i} + (\ddot{\phi}_1 \cos \phi_1 - \dot{\phi}_1^2 \sin \phi_1) \hat{j}]$$

$$\ddot{r}_2 = 2\ddot{r}_1 + \left[-\frac{1}{2}l_2 (\ddot{\phi}_1 + \ddot{\phi}_2) \sin(\phi_1 + \phi_2)\right.$$

$$- (\dot{\phi}_1 + \dot{\phi}_2)^2 \cos(\phi_1 + \phi_2)\Big]\hat{i}$$

$$+ \frac{1}{2}l_2\left[(\ddot{\phi}_1 + \ddot{\phi}_2) \cos(\dot{\phi}_1^2 + \dot{\phi}_1)^2 - (\dot{\phi}_1 + \dot{\phi}_2)^2 \sin(\phi_1 + \phi_2)\right]\hat{j}$$

The dynamic equations are (vector notation)

$$F_1 = m_1\ddot{r}_1$$

$$F_2 = m_2\ddot{r}_2$$

$$\tau_1\hat{k} = I_1\dot{\omega}_1\hat{k} + \omega_1\hat{k}xI_1\omega_1\hat{k}$$

$$\tau_2\hat{k} = I_2\dot{\omega}_2\hat{k} + \omega_2\hat{k}xI_2\omega_2\hat{k} \tag{12.47}$$

Now

$$\omega_1 = \dot{\phi}_1 \qquad \omega_2 = \dot{\phi}_1 + \dot{\phi}_2$$

$$\dot{\omega}_1 = \ddot{\phi}_1 \qquad \dot{\omega}_2 = \ddot{\phi}_1 + \ddot{\phi}_2 \tag{12.48}$$

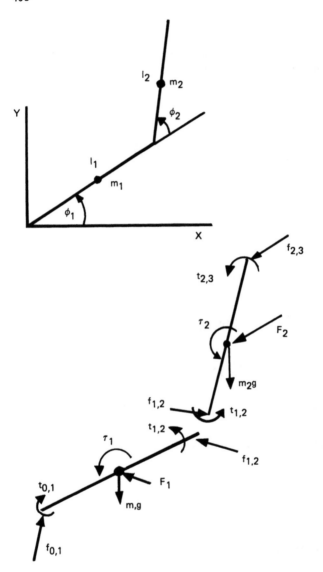

FIGURE 12.9   Formulation of Newton-Euler dynamic equations for a two-link robot.

So

$$\tau_1 = I_1 \ddot{\phi}_1$$

$$\tau_2 = I_2(\ddot{\phi}_1 + \ddot{\phi}_2) \tag{12.49}$$

The force balance equations are (vector notation)

$$F_1 = f_{0,1} - f_{1,2} - m_1 g\hat{j}$$

$$F_2 = f_{1,2} - f_{2,3} - m_2 g\hat{j} \tag{12.50}$$

From these (vector notation)

$$f_{1,2} = F_2 + f_{2,3} + m_2 g\hat{j}$$

$$= m_2 \left\{ [-l_1(\ddot{\phi}_1 \sin \phi_1 + \dot{\phi}_1^{\,2} \cos \phi_1)] \right.$$

$$\left. + \frac{1}{2}l_2[-(\ddot{\phi}_1 + \ddot{\phi}_2) \sin(\phi_1 + \phi_2) - (\dot{\phi}_1 + \dot{\phi}_2)^2 \cos(\phi_1 + \phi_2)] \right\}\hat{i}$$

$$+ m_2 \left\{ [l_1(\ddot{\phi}_1 \cos \phi_1 - \dot{\phi}_1^{\,2} \sin \phi_1)] \right.$$

$$\left. + \frac{1}{2}l_2[(\ddot{\phi}_1 + \ddot{\phi}_2) \cos(\phi_1 + \phi_2) - (\dot{\phi}_1 + \dot{\phi}_2)^2 \sin(\phi_1 + \phi_2) + g] \right\}\hat{j}$$

$$+ f_{2,3x}\hat{i} + f_{2,3y}\hat{j}$$

$$= f_{1,2x}\hat{i} + f_{1,2y}\hat{j} \tag{12.51}$$

$$f_{0,1} = F_1 + f_{1,2} + m_1 g\, \hat{j}$$

$$= \frac{1}{2}m_1 l_1 [-(\ddot{\phi}_1 \sin \phi_1 + \dot{\phi}_1^{\,2} \cos \phi_1)i$$

$$+ (\ddot{\phi}_1 \cos \phi_1 - \dot{\phi}_1^{\,2} \sin \phi_1 + g)j]$$

$$+ m_2 \left\{ l_1(\ddot{\phi}_1 \cos \phi_1 - \dot{\phi}_1^2 \sin \phi_1) \right.$$

$$+ \frac{1}{2}l_2[(\ddot{\phi}_1 + \ddot{\phi}_2) \cos(\phi_1 + \phi_2) - (\dot{\phi}_1 + \ddot{\phi}_2)^2 \sin(\dot{\phi}_1 + \phi_2)] + g \left. \right\} \hat{j}$$

$$+ f_{2,3x}\hat{i} + f_{2,y}\hat{j}$$

$$= f_{0,1x}\hat{i} + f_{0,1y}\hat{j} \qquad\qquad (12.52)$$

For torque balance

$$\tau_2\hat{k} = t_{1,2}\hat{k} - t_{2,3}\hat{k} + r_2^* \times f_{1,2} + r_2^* \times f_{2,3}$$

$$\tau_1\hat{k} = t_{0,1}\hat{k} - t_{1,2}\hat{k} + r_1^* \times f_{0,1} + r_1^* \times f_{1,2} \qquad\qquad (12.53)$$

where

$$r_1^* = -\frac{1}{2}l_1(\cos \phi_1\hat{i} + \sin \phi_1\hat{j})$$

$$r_2^* = -\frac{1}{2}l_2[\cos(\phi_1 + \phi_2)\hat{i} + \sin(\phi_1 + \phi_2)\hat{j}]$$

Solving these for $t_{1,2}$ and $t_{0,1}$, we obtain

$$t_{1,2}\hat{k} = \tau_2\hat{k} + t_{2,3}\hat{k} - r_2^* \times f_{1,2} + r_2^* \times f_{2,3}$$

$$t_{0,1}\hat{k} = \tau_1\hat{k} + t_{1,2}\hat{k} - r_1^* \times f_{0,1} - r_1^* \times f_{1,2} \qquad\qquad (12.54)$$

By substituting and simplifying,

$$t_{1,2}\hat{k} = I_2(\ddot{\phi}_1 + \ddot{\phi}_2)\hat{k} + t_{2,3}\hat{k}$$

$$+ \frac{1}{2}l_2[\cos(\phi_1 + \phi_2)\hat{i} + \sin(\phi_1 + \phi_2)\hat{j}] \times [(f_{1,2x} + f_{2,3x})\hat{i}$$

$$+ (f_{1,2y} + f_{2,3y})\hat{j} \qquad\qquad (12.55)$$

Finally

$$t_{1,2} = \left\{ I_2 + \frac{1}{2}m_2l_1l_2 \cos \phi_2 + \frac{1}{4}m_2l_2^2 \right\} \ddot{\phi}_1 + \left\{ I_2 + \frac{1}{4}m_2l_2^2 \right\} \ddot{\phi}_2$$

$$+ \frac{1}{2} m_2 l_1 l_2 \dot{\phi}_1^2 \sin \phi_2 + \frac{1}{2} m_2 l_2 g \cos(\phi_1 + \phi_2)$$

$$-l_2 \sin(\phi_1 + \phi_2) f_{2,3x} + l_2 \cos(\phi_1 + \phi_2) f_{2,3y} \qquad (12.56)$$

$$t_{0,1} = \left[I_1 + I_2 + m_2 l_1 l_2 \cos \phi_2 + \frac{1}{4}(m_2 l_2^2 + m_1 l_1^2) + m_2 l_1^2 \right] \ddot{\phi}_1$$

$$+ \left(I_2 + \frac{1}{4} m_2 l_2^2 + \frac{1}{2} m_2 l_1 l_2 \cos \phi_2 \right) \ddot{\phi}_2$$

$$- m_2 l_1 l_2 \dot{\phi}_1 \dot{\phi}_2 \sin \phi_2 - \frac{1}{2} m_2 l_1 l_2 \dot{\phi}_2^2 \sin \phi_2$$

$$- [l_1 \sin \phi_1 + l_2 \sin(\phi_1 + \phi_2)] f_{2,3x}$$

$$+ [l_1 \cos \phi_1 + l_2 \cos(\phi_1 + \phi_2)] f_{2,3y}$$

$$+ \left[\frac{m_2 l_2}{2} \cos(\phi_1 \phi_2) + l_1 \left(\frac{m_1}{2} + m_2 \right) \cos \phi_1 \right] g + t_{2,3} \qquad (12.57)$$

In this section we will turn our attention to the control of manipulator motions. The joint actuators of a manipulator must be controlled in a specific manner to accomplish a given task. The control of a manipulator requires a fundamental understanding of the basics of control theory. Therefore we will first present the derivation of the transfer function of electric servomotors and hydraulic servomotors normally found in robot systems. We will then discuss the transfer function of a single manipulator joint actuator system consisting on joint actuator and link inertias. The performance aspects of control systems in robots are treated next by considering the effects of damping, multiple joint coupling, and steady-state errors. This section is concluded by a discussion of some enhanced control features such as adaptive control for robot systems.

12.6   TRANSFER FUNCTION OF MANIPULATOR DRIVES

Electric Servomotors

The governing equations for an armature-controlled DC motor are obtained as follows. The driving torque of the motor τ is proportional to the current applied to the motor armature I_a:

$$\tau = K_t I_a \qquad (12.58)$$

where K_t is the torque constant. The electrical characteristics of the armature are represented by

$$V_a = R_a I_a + V_b + L_a \frac{dI_a}{dt} \qquad (12.59)$$

where V_a is the voltage applied to the armature and V_b is the back EMF. The inductance of the armature is usually assumed to be small enough to be neglected. The back EMF, which is proportional to the angualr velocity of the motor shaft, is represented by

$$V_b = K_b \dot{\theta} \qquad (12.60)$$

The input torque is obtained directly from these equations as

$$\tau = \frac{K_t}{R_a}\left(V_a - K_b \dot{\theta} - L_a \frac{dI_a}{dt}\right) \qquad (12.61)$$

For a motor driving a load with an effective inertia J and viscous friction factor f, the equation of motion is

$$\tau = J\ddot{\theta} + f\dot{\theta} \qquad (12.62)$$

From the preceding two equations, assuming the inductance is negligible,

$$J\ddot{\theta} + \left(f + \frac{K_b K_t}{R_a}\right)\dot{\theta} = \frac{K_t V_a}{R_a} \qquad (12.63)$$

or

$$J\ddot{\theta} + F\dot{\theta} = K_m V_a$$

where $F = f + K_b K_t/R$ is the effective viscous friction coefficient and $K_m = K_t/R_a$ is the effective motor gain.

The transfer function of the servomotor is thus

$$\frac{\theta(s)}{V(s)} = \frac{K_m}{s(Js + F)} \qquad (12.64)$$

The corresponding block diagrams are shown in Figure 12.10.

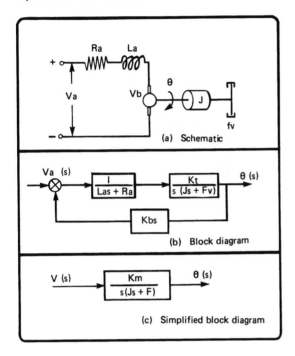

(a) Schematic

(b) Block diagram

(c) Simplified block diagram

FIGURE 12.10   Transfer function of an armature-controlled DC motor.

Hydraulic Servomotors

In robots driven by hydraulic servomotors, the medium power transmission is a fluid such as oil under pressure. Pneumatic systems, on the other hand, generally use air as the medium for power transmission. A schematic of a basic hydraulic system and the common arrangement of the valve-actuator unit in hydraulic drives are shown in Figure 12.11. To derive the transfer function, let us assume incompressible flow and ignore leakage around the pistons. The flow rate equation is

$$A \frac{dx_A}{dt} = K_1 x_v - K_2 P_D \tag{12.65}$$

where P_D is the pressure difference and K_1 and K_2 are constant parameters which are typically obtained from linearized valve characteristics; K_1 is the vertical distance between constant-pressure lines divided by the change in stem motion, and K_2 is the slope of a typical characteristic line.

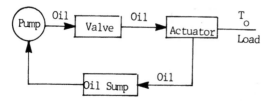

(a) Components of a hydraulic drive system

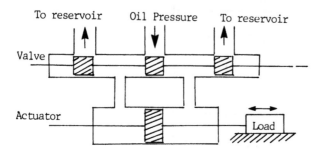

(b) Hydraulic valve

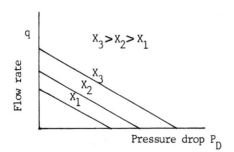

(c) Linearized valve characteristics

FIGURE 12.11   Hydraulic control system.

The equation of motion of the piston is given by

$$M \frac{d^2 x_A}{dt^2} + f \frac{dx_A}{dt} = AP_D \qquad (12.66)$$

Combining these two equations yields

$$M \frac{d^2 x_A}{dt^2} + \left(f + \frac{A^2}{K_2} \right) \frac{dx_A}{dt} = A \left(\frac{K_1}{K_2} \right) x_V \qquad (12.67)$$

The transfer function is then obtained in the notation of the Laplace transform as

$$\frac{X_A(s)}{X_V(s)} = \frac{A(K_1/K_2)}{Ms^2 + (f + A^2/K_2)s} \qquad (12.68)$$

12.7   BASIC MANIPULATOR CONTROL

The basic dynamic equation of a joint actuator of a manipulator with a desired joint motion θ_d can be written as

$$J\ddot{\theta} + F\dot{\theta} = k_a \dot{\theta}_d \qquad (12.69)$$

where k_a is the motor gain factor, F the equivalent viscous friction factor, and J the effective inertia at the actuator. The effective inertia is normally found by combining the link inertia and the actuator inertia in the form

$$J = I_1 + Ir^2 \qquad (12.70)$$

where I_1 is the link inertia, I the actuator inertia, and r the reduction gear ratio between the actuator and the joint drive. The transfer function of an actuator is derived from Equation (12.69) as

$$\frac{s\theta(s)}{s\theta_d(s)} = \frac{K_a}{sJ + F} \qquad (12.71)$$

If rate feedback from a tachometer generator (Figure 12.12a) is used, the resulting dynamic equation and transfer function are, respectively,

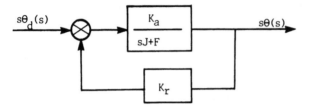

(a) Rate feedback

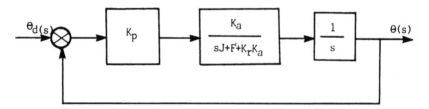

(b) Proportional plus rate control

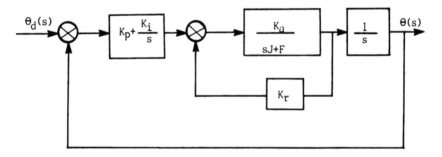

(c) Proportional, integral and rate control

FIGURE 12.12   Basic control of a manipulator.

$$J\ddot{\theta} + (F + k_r k_a)\dot{\theta} = k_a \dot{\theta}_d \tag{12.72}$$

$$\frac{s\theta(s)}{s\theta_d(s)} = \frac{K_a}{sJ + (F + K_r K_a)} \tag{12.73}$$

With an additional position feedback (Figure 12.12b) the dynamic equation becomes

$$J\ddot{\theta} + (F + k_r k_a)\dot{\theta} + k_a k_p \theta = k_a k_p \theta_d \tag{12.74}$$

This yields the transfer function as

$$\frac{\theta(s)}{\theta_d(s)} = \frac{k_p k_a}{s^2 + s(F + k_r k_a) + k_r k_p} \tag{12.75}$$

Figure 12.12c shows a block diagram of a typical position servo for a manipulator link. This servo has combination of proportional, intergral, and derivative control. The transfer function is

$$\frac{\theta(s)}{\theta_i(s)} = \frac{k_a(k_p s + k_i)}{s^3 J + (F + k_a k_r)s^2 + k_p k_a s + k_i k_a} \tag{12.76}$$

Usually in selecting gains for acceptable performance of the position servo, the integral factor k_i is assumed to be zero. With this assumption, the position servo is reduced to an equivalent second-order system which can be used to study its performance. The transfer function of the equivalent second-order system is

$$\frac{\theta(s)}{\theta_i(s)} = \frac{k_a k_p}{s^2 J + (F + k_a k_r)s + k_p k_a} \tag{12.77}$$

The characteristic frequency ω_o and damping ratio ξ for this second-order model are

$$\omega_0^2 = \frac{k_p k_a}{J} \tag{12.78}$$

$$\xi = \frac{F + k_a k_r}{2\sqrt{Jk_p k_a}} \tag{12.79}$$

12.8 IMPROVEMENT OF CONTROL PERFORMANCE

Damping

In manipulator motions, overshoot is not permissable in the step input response. In insertion operations, for instance, overhsoot would cause the manipulator hand to travel beyond the required location. Overshoot is eliminated by choosing the damping factor to reduce the system to an overdamped second order.

The condition for an overdamped system ($\phi > 1$) is obtained from Equation (12.79) as

$$F + k_r k_a > 2\sqrt{Jk_p k_a}$$

Fixed-position servo gain and variable-velocity servo gain are usually appropriate for a manipulator link which essentially has a variable inertia. For critical damping ($\xi = 1$) the velocity gain can be obtained as

$$k_r = \frac{1}{k_a}(2\sqrt{Jk_p k_a} - F) \qquad (12.80)$$

For a pair of known values (k_r^*, J^*) the velocity gain k_r for any J is obtained from

$$\frac{k_r k_a + F}{\sqrt{J}} = \frac{k_r^* k_a + F}{\sqrt{J^*}} \qquad (12.81)$$

as

$$k_r = \frac{1}{k_a}(G\sqrt{J} - F)$$

where

$$G = \frac{k_r^* k_a + F}{\sqrt{J^*}}$$

Compensation for Steady-State Errors

In manipulator operations, it is desirable to improve the performance by reducing the steady-state positional errors due to an external torque disturbance and static and dynamic Coulomb friction. The steady-state error due to Coulomb friction gives a measure of the repeatability of the joint. The static Coulomb friction T_s is the frictional effect which must be overcome before joint motion occurs. When the joint is in motion this torque drops to the dynamic Coulomb friction T_d, which opposes the motion. The repeatability is defined as the position error which causes a servo torque response equal to T_d. Poor repeatability may be improved by a feedforward torque to the joint given by

$$T_{ff}(S) = \begin{cases} \dfrac{T_d}{S} & \text{for } \dot{\theta} > 0 \\[3mm] -\dfrac{T_d}{S} & \text{for } \dot{\theta} < 0 \end{cases} \qquad (12.82)$$

When the joint is stationary, the necessary impulsive correction torque is given by

$$
T_{ff}(S) = \begin{cases} T_S & \text{for } \theta_e > 0 \\[2mm] -T_S & \text{for } \theta_e < 0 \end{cases}
\tag{12.83}
$$

Gravity compensation is achieved by providing an additional feed-forward torque to the joint servos. This torque may be set equal to the gravity loading torque term in the dynamics of the manipulator.

For manipulators used in moving coordinate systems, such as when an end effector is tracking a conveyer, the steady-state velocity error is a significant control parameter. This error can be eliminated by providing feedforward based on the desired velocity.

Acceleration errors are relatively unimportant at high speeds of the manipulator but become quite significant at the start (liftoff) and finish (setdown) phases of the motion. Acceleration errors are compensated by including a feedforward effective inertia gain. The effective inertia may be approximated by its minimum value to provide partial compensation. However, if the maximum value is used, an undesirable overhsooting motion may result.

Joint Coupling Effects

The control schemes described above are for a single link of the manipulator considered separately and independent of the other links. When the motions of all the links are considered simultaneously three other effects are introduced in the dynamic equations. These are the inertial coupling, centripetal, and Coriolis dynamic terms. The inertial coupling is accounted for by an additional feedforward gain factor to the appropriate joint servo.

The centripetal and Coriolis effects are often significant at high-speed motions of the manipulator. Although these do not affect the stability, they cause position errors. These effects are routinely ignored when using the schemes in Figure 12.12 for control of simple manipulators where positional accuracy may be un-inmportant an high speeds.

In real-time control of manipulators, efficient calculation of the joint torque is extremely important. Many schemes can be used to achieve on-line computational efficiency of computed torque control. First, the basic equation for the manipulator may be simplified by eliminating the Coriolis and centrifugal terms as insignificant. This assumption causes poor manipulator performance when velocities are large. The second approach is to replace some calculations by table lookup schemes, in which the manipulator dynamics is precomputed and stored. Finally, recursive formulations based on Lagrangian and Newton-Euler methods may be used. In the recursive methods and velocities and accelerations are found sequentially, starting from the base link or end effector of the manipulator.

In assembly operations manipulators may be required to exert forces and control position. To do this the position servo must be changed to a joint torque servo, which is normally configured in the form of a position servo with strain as the error signal. The error is converted to a torque by a spring gain constant k_s. The torque gain k_t then amplifies the error torque. There is a feedforward term from the joint torque to the motor because the joint is connected directly to the motor. To obtain the performance parameters of a torque servo, it is assumed that the end of the joint is fixed and cannot move. This simplifies the inertia of the system to that of the drive system referred to the output shaft.

Adaptive Control

Adaptive control refers to a broad class of control systems which adjust the control signal to accommodate changes in the system dynamics and environmental disturbances. This method of control is well suited to the manipulator control problem, which is highly nonlinear. The major nonlinear effects that limit the dynamic performance of industrial robots are the dynamic effects caused by variable payload and the changing configuration of the robot in the workspace. Because of these nonlinearities, classical linear control systems do not delivery uniformly high performance over a wide range of system operating conditions. Model reference adaptive control schemes compensate for both the nonlinear dynamics of the system and the degradation caused by an unknown paylaod.

We illustrate the control features of a manipulator by considering an example used in robot applications (see Figure 12.13). Several enhanced control options improve the efficiency of deburring operations with tools such as rotary files. One of these options is servo soft, which enables the desired tool contact pressure to be maintained at any point in the program. The contract pressure is not affected by tool wear or variations in part shape. Another option is adaptive control, which enables the robot to select appropriate feed rates for various contact conditions. A fast feedback is used until the tool encounters a burr, when a slower feed rate is used. The slower feed rate used for deburring reduces excessive wear and deflection of the tool.

In polishing complex shapes, soft servo provides the contact pressures for optimum results. Adaptive control with a suitable sensor enables parts of uncertain shape and indefinite positions to be efficiently polished. Adaptive control also compensates for wear of the brush.

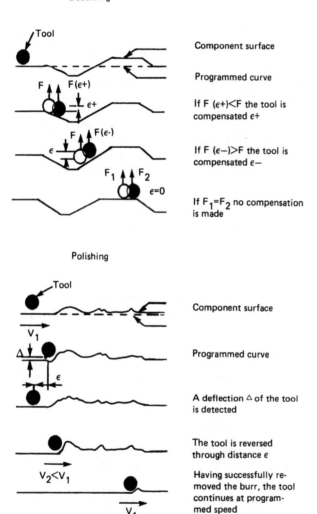

Deburring

Tool

Component surface

Programmed curve

If F (ε+)<F the tool is
compensated ε+

If F (ε−)>F the tool is
compensated ε−

If F_1=F_2 no compensation
is made

Polishing

Tool

Component surface

Programmed curve

A deflection △ of the tool
is detected

The tool is reversed
through distance ε

Having successfully re-
moved the burr, the tool
continues at program-
med speed

FIGURE 12.13   Adaptive control for deburring and polishing.
(Courtesy of ASEA, Inc.)

REVIEW QUESTIONS

1. Using the Lagrangian method, derive the dynamic equations of
    the four geometric types of robots. Use an appropriate dynamic
    model of three degrees of freedom including masses of motors,
    links payload, and distributed mass effects. Show the contribu-
    tion of these factors in the joint forces and torques required.
2. Rewrite the dynamic equations derived in problem 1, using the
    concept of dynamic coefficients.
3. Describe two different kinematic plans trajectories that can be
    used for the joint motion of a robot. Using these kinematic
    motions, evaluate numerically the joint torques and forces over
    a period of joint coordinated movements. Plot your results on
    graph paper.
4. Develop an interactive computer program to study the dynamics
    of industrial robots. Include subroutines to generate the plots
    on a computer. Use a number of different dynamic models.

FURTHER READING

Albus, J. S., "Data Storage in the Cerebellar Model Articulation
    Controller," *Trans. ASME, J. Dynamic Syst. Meas. Control.*,
    Vol. 97, Sept. 1975, pp. 228–233.

Albus, J. S., "A New Approach to Manipulator Controls: The
    Cerebellar Model Articulation Controller (CMAC)," *Trans. ASME,
    J. Dynamic Syst. Meas. Control*, Vol. 97, Sept. 1975, pp. 220–
    227.

Bejczy, A. K., " Robot Arm Dynamicas and Control," NASA-Jet
    Propulsion Laboratory Tech. Memo. 33-669, Feb. 1974.

Binford, T., "Exploratory Studies of Computer Integrated Assembly
    Systems," NSF Progress Report, Stanford Artificial Intelligence
    Laboratory Memo AIM-285, July 1976.

Groome, R. C., "Force Feedback Steering of a Teleoperator System,"
    S.M. Thesis, MIT, 1972.

Hollerbach, J. M., "A Recursive Lagrangian Formulation of Manipu-
    lator Dynamics and a Comparative Study of Dynamics Formulation,"
    IEEE Trans. Syst. Man. Cyber., Vol. SMC-10, No. 11, Nov.
    1980, pp. 730–736.

Horn, B. K. P. and Raibert, M. H., "Configuration Space Control,"
    Industrial Robot, June 1978, pp. 69–73.

Horn, B. K. P., Hirokawa, K., and Vazirani, V., "Dynamics of a
    Three Degree of Freedom Kinematic Chain," Stanford Artificial
    Intelligence Project, Memo No. 478, Oct. 1977.

Kahn, M. and Roth, B., "The Near-Minimum-Time Control of Open
   Loop Kinematic Chains," *Trans ASME, Ser. G*, Vol. 93, 1971,
   pp. 164–172.

Lewis, R. A., "Autonomous Manipulations on a Robot: Summary of
   Manipulator Software Functions," Jet Propulsion Laboratory
   Tech. Memo. TM 33-679, 1974.

Luh, J., Walker, M., and Pual, R., "On-Line Computational Scheme
   for Mechanical Manipulation," *Trans. ASME, J. Dynamic Syst.
   Meas. Control*, Vol. 102, No. 2, June 1980, pp. 69–71.

Luh, J., Walker, M., and Paul, R., "Resolved Acceleration Control
   of Mechanical Manipulation," *IEEE Trans. Autom. Control*,
   Vol. AC-25, No. 3, June 1980, pp. 468–474.

Orin, D. D., McGhee, R. B., and Vukobratovic, M., "Kinematic
   and Kinetic Analysis of Open-Chain Linkages Utilizing Newton-
   Euler Methods," *Math. Biosci.*, Vol. 43, 1979, pp. 107–130.

Paul, R., "Advanced Industrial Robot Control Systems," First
   Report, Purdue University Memo EE 78-25, May 1978.

Paul, R. P., "Kinematic Control Equations for Manipulators," *IEEE
   Trans. Syst. Man. Cyber.*, 1981.

Paul, R. P., "Modeling, Trajectory Calculation and Servoing of a
   Computer Controlled Arm," Stanford Artificial Intelligence
   Laboratory, Stanford University, AIM 177, 1972.

Pieper, D. L., "The Kinematics of Manipulators under Computer
   Control," Stanford Artificial Intelligence Laboratory, Stanford
   University, AIM 72, 1968.

Raibert, M. H. and Horn, B. K. P., "Manipulator Control Using the
   Configuration Space Method," Industrial Robot, Vol. 5, No. 2,
   June, 1978, pp.69-73.

Roderic, M. D., "The Discrete Control of a Manipulator," Stanford
   Artificial Intelligence Laboratory, Stanford University, AIM 287,
   Aut. 1976.

Rosen, C. and Nitzan, D., "Exploratory Research in Advanced
   Automation," Second Report, Stanford Research Institute,
   Aug. 1974.

Roth, B., "Performance Evaluation of Manipulators from a Kinematic
   Viewpoint," National Bureau of Standards Report 459, 1976.

Scheinman, V. D., "Design of a Computer Manipulator," Stanford
   Artificial Intelligence Laboratory, Stanford University, AIM 92,
   1969.

Taylor, R. H., "Planning and Execution of Straight-Line
Manipulator Trajectories," IBM Research Report RC 6657,
July 1977.

Uicker, J. J., Jr., "Dynamic Force Analysis of Spatial Linkages,"
Journal of Applied Mechanics, *Trans ASME*, June 1967.

Wu, C.-H. and Paul, R. P., "Manipulator Compliance Based on Joint
Torque Control," *19th IEEE Conf. Decision Control*, Dec. 1980,
pp. 88-94.

Index

Milton Keynes UK
Ingram Content Group UK Ltd.
UKHW021844071024
449327UK00021B/1539